DATE DUE

DEC 05 1998			
NOV 05 2000			
MAY 17 2001			

HIGHSMITH 45-220

Super-High-Definition Images

Beyond HDTV

For a complete listing of the *Artech House Telecommunications Library,* turn to the back of this book.

Super-High-Definition Images

Beyond HDTV

Sadayasu Ono
Naohisa Ohta
Tomonori Aoyama

Artech House
Boston • London

Library of Congress Cataloging-in-Publication Data

Ohta, Naohisa.
    Super-high definition images: beyond HDTV / Naohisa Ohta, Sadayasu Ono, and
Tomonori Aoyama.
      p. cm.
    Includes bibliographical references and index.
    ISBN 0-89006-674-4 (hard : acid-free paper)
      1. Image processing—Digital techniques.   2. Multimedia systems.   3. Digital tele-
vision.   I. Ono, Sadayasu.   II. Aoyama, Tomonori.   III. Title
    TA1637.O37   1995   94-49709
    621.36'7—dc20   CIP

British Library Cataloguing in Publication Data

Ohta, Naohisa
Super-high-definition Images: Beyond HDTV
I. Title
621.388

ISBN 0-89006-674-4

© 1995 ARTECH HOUSE, INC.
685 Canton Street
Norwood, MA 02062

International Standard Book Number: 0-89006-674-4
Library of Congress Catalog Card Number: 94-49709

10 9 8 7 6 5 4 3 2 1

Contents

Preface

Super-high-definition (SHD) imaging is a platform being advanced to support media integration. More simply, the aim of SHD is to support all kinds of pre-existing media, both still and motion, and to provide a level of image that will satisfy all quality requirements. The development of the broadband integrated services digital network (B-ISDN) is ushering in a new media revolution by allowing the transfer of this SHD medium in real time and at affordable rates.

SHD encompasses high-definition television (HDTV), for which the Federal Communications Commission (FCC) is currently standardizing pixels resolution and frame rate. Since both forms of media are digital, they are highly compatible. It is not a coincidence that SHD and HDTV are all-digital methods of data transfer, but rather an inevitable consequence of the trend toward digitalization in media. The goal of SHD imaging is media integration, or, more precisely, enabling the mutual exchange of image information between all types of media. The only way to accomplish this while maintaining a high level of image quality is by using digital technology.

Are special coding algorithms required for compressing SHD images? This book will show that SHD images can be transferred over B-ISDN at the standard rate of 150 Mbps by compressing them using MPEG-2 (a coding algorithm that is nearing standardization and is also used for HDTV). The authors expect that enhancing the coding method used for digital HDTV will allow SHD images to be coded, transferred, and stored at an acceptable level of quality. And since these are digital images, they are compatible with HDTV, no matter what kind of coding algorithm is used.

The use of coding results in a good deal of trial and error. It is therefore vital to provide flexibility by implementing the CODEC as a digital signal processor (DSP). Some fields more than others place extremely heavy demands on SHD imaging, and this makes achieving flexibility even more important. This flexibility is indispensable for realizing the many objectives required in the development of multimedia and hypermedia.

The authors believe that SHD imaging can be used for existing multimedia communication platforms, but accurately defining existing multimedia communication platforms is quite a difficult task. This book will not directly address this issue; rather, it will try to approach a solution by describing a number of applications for SHD.

The structure of this book is as follows. Chapter 1 discusses the general circumstances regarding image media and the significance of SHD imaging. Chapter 2 describes media integration, which is the aim of SDH imaging. In addition, Chapter 2 describes the level of quality that SDH must achieve to obtain media integration, which adds to the definition of SHD imaging in this book. Chapter 3 describes the results of using basic algorithms to compress images and JPEG and MPEG-2 to compress SHD images (both still and moving). Chapter 4 introduces the current state of image input devices and describes the types of computers that can be used for SHD image processing systems. Chapter 5 discusses the technical requirements for systems that can process moving SHD images in real time. In particular, Chapter 5 describes a parallel processing DSP system, which is the most promising method of achieving an architecture that is both programmable and capable of handling SHD images, and this chapter also introduces the experimental parallel processing system called NOVI-II. Chapter 6 discusses image display systems, and Chapter 7 gives an overview of the experimental SHD processing system being used at the authors' laboratory. Chapter 8 describes, from various angles, future applications of SHD imaging.

Super-High-Definition Images: Beyond HDTV does not contain an in-depth discussion of established technology. Rather, by making proposals regarding future new image media, this book attempts to raise the reader's awareness of SHD imaging in an effort to stimulate developments in this field.

Foreword

Advances in electronics technology and the growth of broad-band communication networks have stimulated the demand for a higher definition image media. To achieve the resolution required by publishers, CAD users, doctors, scientists, and entertainment studios, a color TV system superior to HDTV is necessary. This new image category is referred to as super-high-definition (SHD) images. It is expected that this new kind of image media will subsume all existing image systems and bring us more realistic visual and multimedia communications.

This up-to-date book on SHD images provides comprehensive discussions of current and future media integration, multimedia communications, and the supporting technologies. It discusses why all-digital SHD images will become the key to media integration, describing how SHD images are different from existing image format and what kind of new applications will arise. The necessary technologies for achieving SHD image processing are discussed. The main discussions are capture/display technology, compression algorithms, and signal processing technologies. Experimental systems for SHD image processing are also introduced, focusing on parallel DSP systems. Finally, the authors' view of SHD image applications beyond HDTV towards the multimedia/hypermedia age is presented.

Introduction 1

It is said that the multimedia age is coming in the near future. It is difficult to describe exactly what, in fact, the term *multimedia* means. A bare-bones description might be that it simultaneously encompasses voice, audio, text, and images and will be used by people in their intellectual pursuits. *Hypermedia* is another term that is frequently heard. It refers to the creation of natural links between various forms of media, which allows people to pursue their intellectual activities while creating these links. The truth is, however, that multimedia and hypermedia are still not well understood.

A currently popular environment in which multimedia is being used is the personal computer (PC) connected to a communication line. Among older forms of image media, one form that has the most influence (the amount depends on the country) is television. Even entertainment video, including movies and TV shows, through digitization, is approaching the multimedia age. For example, video on demand (VOD), a typical interactive video service through cable TV or communication networks is now seen as the major source of revenues from the industry. Likewise, the Federal Communications Commission's (FCC) move to standardize advanced TV (ATV) and, in reaction to that, Europe's trend towards high-definition TV (HDTV) are probably the first steps towards changing the existing TV technology—which has been limited to one-way information transmission—into a far more exciting form of media [1].

The purpose of super high definition (SHD), the subject of this book, is not merely to provide high image quality, but, more importantly, to develop a platform for the integration of image media in the coming multimedia/hypermedia age. The requirements for a platform that can integrate image media are the following. It must be able to both handle existing image media without any problems related to image quality and be easy to use. It must also encompass digital HDTV standards. In this respect, SHD is a likely candidate for the technology that reaches beyond HDTV. (We will pursue this topic in Chapter 2). In this chapter we will describe from many angles (primarily the digitization of image media and communication networks and the development of computers) the background of the coming multimedia age and the technological background that led to the development of SHD imaging.

1.1 TECHNOLOGICAL BACKGROUND

The technological factors that will bring about the multimedia age can be broken down into three elements: (1) digitization of image media, (2) digitization of communication networks, and (3) advances in computers. These three elements, of course, could not have been realized without advances in electronic technology, particularly in very-large-scale integration (VLSI) technology, and developments in digital communication technology, primarily in the area of fiber optics. In this book, however, electronic and digital communication technology will not be discussed; instead we discuss the technological background from the three angles described above.

1.1.1 Digitization of Image Media

The trend towards digitizing image media is advancing in all fields. In the field of prepackaged audio media, compact discs (CD) now far outnumber analog disks (records). It is common knowledge that audio CDs store audio signals digitally. In the field of image media, video cassettes are still the dominate form of prepackaged media. However, thanks to modern compression technology such as the Joint Photographic Experts Group (JPEG) [2] and the Motion Picture Experts Group (MPEG) [3] methods, the digitization of image media has rapidly advanced in areas such as CD-ROM (read-only memory). Especially deserving attention is the use of CD-ROM products on Macintoshes, PCs, and workstations to playback digitally compressed still and moving images. The trend of digitizing image media is also advancing in conventional industry, such as printing, video editing, and broadcasting.

Various forms of media integration can clearly be seen in the circumstances described above. Some examples of these include digital video interactive (DVI) [4], desktop publishing, and hypermedia PCs. Along with the trend of packaged media going digital, the need will arise for services that use digital communication networks. In the United States, Video Dial Tone [5] is an example of such an experiment. Broadcast media has also begun to be digitized. In the United States, compression products are establishing MPEG2 [6] as the de facto standard, even for ATV. Digital HDTV in Europe is likely to follow suit.

1.1.2 Digitization of Communication Networks

With the help of VLSI and fiber optics, communication networks too are advancing from mere 3.4-kHz telephone line networks to those offering integrated services digital network (ISDN) [7] and even broadband ISDN (B-ISDN). The basic speed on ISDN, often referred to as a *B channel*, is 64 Kbps, and on B-ISDN the figure is about 150 Mbps. Digitization increases the rate of data transmission. Another important point is that all kinds of digital data, regardless of media type, can be sent and received over the same line, making it possible to integrate services. This is where the abbreviation ISDN comes from.

Building digital communication networks and increasing basic transfer rates raises the image quality of media that can be transferred. Existing networks using 3.4-kHz telephone lines are capable of transmitting facsimiles and slow scans. At ISDN's data rates ranging from 64 Kbps to 1.5 Mbps (2 Mbps in Europe), it is possible to compress and transfer images with a resolution comparable to high-quality faxes and broadcast TV. The video dial tone service is also of this level of quality. Transfer speeds, however, are still too slow to transfer images at their original resolution. This has led to experiments with variable-rate transfers that ingeniously use the temporal redundancy of image sequences [8]. In the near future, when B-ISDN comes to be widely used, users will be offered direct fiber-optic connections to their homes, which will enable communication at 150 Mbps. This speed will allow the transfer of entertainment video and broadcast TV at an acceptable level without any compression. It will also make possible the transmission of digital ATV. There are, however, nontechnical social issues that remain to be solved, namely the integration of broadcast and communications.

The above discussion comes from the viewpoint of network providers. It is also important to see consumers' attitudes concerning the use of digital networks. According to a survey [9], consumers are much more interested in using digital networks for information access, community involvement, self-improvement, and communications for all kinds of intellectual activities than for entertainment. This implies that some kind of professional use of multimedia/hypermedia other than entertainment will play an important role in the future development of digital networks.

1.1.3 Development of Computers

Advances in VLSI technology have increased semiconductor memory speed, mass production rates, and processor speed, resulting in a remarkable increase in computer compactness and performance. Current PCs have the same calculating power, but only 1/100 the size of early mainframe computers. For example, the world's first supercomputer, the CDC6600 developed in 1964, had a computing power of 1 Mflops, while today's average workstation has a 10-Mips central processing unit (CPU). PCs have already surpassed early supercomputers in the amount of semiconductor memory used. Also, to be able to process image signals, many PCs and workstations use a dedicated digital signal processor (DSP) as a coprocessor. DSPs can operate at 10 to 40 Mflops and have enabled small PCs configured as multimedia terminals to surpass the capabilities of past supercomputers.

These small-sized computers with their advanced calculating capabilities and abundant memory are becoming an important tool for integration. Desktop publishing, in the form of printing images and text received over communication lines, is possible by combining a high-resolution color printer with a PC that has a board that can execute JPEG algorithms (a standardized method of compressing still images).

A conceptual view of the relationship between the three technological backgrounds described earlier and multimedia/hypermedia is shown in Figure 1.1. Take particular notice of the image media in this figure. It is apparent that a multimedia/hypermedia environment can be created by combining B-ISDN, digitized image media, and high-performance, small-sized computers.

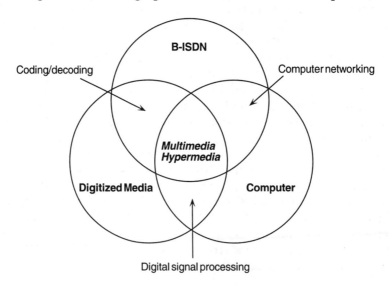

Figure 1.1 Conceptual view of the relationship between multimedia/hypermedia and technological backgrounds.

1.2 REQUIREMENTS FOR SUPER-HIGH-DEFINITION DIGITAL IMAGES

The question of what are the image media requirements in the multimedia/hypermedia age needs to be addressed. Is it enough to simply combine text, pictures, and TV, which have been developing independently? Exactly how image media will be used in the multimedia/hypermedia age is not clear, but what can be said is that it is at least vital to create a platform that can handle existing image media with next to no loss in quality. It is obvious that such a platform must be digital because digital processing is the only way to achieve compatibility at a high-quality level between different forms of media. Also, developments in electronics, primarily in VLSI technology, are making it technically feasible to develop such digital platforms. In the final analysis, it is the increased computer and DSP capabilities along with the high-quality image input, output, and display systems that have made possible the development of these sorts of platforms.

What do telecommunication services have to do to be able to interactively send and receive digital media over communication networks? The ultimate goal of telecommunication services is to emulate as closely as possible the feeling of speaking face to face between people who are actually physically separated. In the past there were a number of barriers in achieving this, but as described in Section 1.1, these technical barriers are close to being overcome. In the business world, telephone conferencing is considered indispensable and as a result is widely used. Video conferencing is still rather rare because of its low value compared to high communication costs. We feel the main factor here in the paucity of video conferencing is its poor image quality, or, to put it simply, incomplete, low-quality images are next to worthless. This situation will change completely if one is able to see on a normal monitor (without a zoom function) subtle changes in facial expression, to display document titles without showing the content, and, when desired, to show and/or view text on a U.S. standard letter-size page or a Japanese/European standard A4 page. One key requirement is that the service be offered at a reasonable cost.

Transferring high-definition images requires a 150-Mbps line such as that offered by B-ISDN. Conversely, the desire to transmit such images is a motivating factor towards installing fiber-optic cable. Among the technological developments that will allow the sending and receiving of high-quality image media in the near future, the spread of B-ISDN is extremely important [10].

1.3 SUMMARY

As we have described in this chapter, SHD imaging was conceived as a platform to integrate the media essential for the multimedia/hypermedia age, and has advanced naturally due to developments in electronic technology. At the same time, this platform will develop high-end services that will hasten the spread of B-ISDN. While HDTV and digital HDTV (ATV in the United States) are ultimately just an extension of broadcast TV, SHD imaging has the higher goal of media integration, including communication. This book will introduce the trends in technology for and application of SHD imaging, primarily the current state of research on SHD image technology that is being conducted at our laboratories. Rather than describing the established technology, this book places more emphasis on raising people's awareness of SHD imaging. In this book, however, discussion of media integration will be limited to images and will not deal with voice and audio, because compared to images, there are few technical difficulties in integrating these two types of sound media.

References

[1]   Gilder, G., *Life After Television*, New York: W.W. Norton & Company, 1992.

[2]   Pennebaker, W.B., and J.L. Mitchell, *JPEG Still Image Data Compression Standard*, New York: Van Nostrand Reinhold, 1993.

[3]   LeGall, D., "MPEG: A Video Compression Standard for Multimedia Applications," *Communications of the ACM*, Vol. 34, No. 4, April 1991, pp. 46–58.

[4]   Fox, E.A., "Advances in Interactive Digital Multimedia Systems," *Computer*, Vol. 24, No. 10, 1991, pp. 9–21.

[5]   Hsing, T.R., C.-T. Chen, and J.A. Bellisio, "Video Communications and Services in the Copper Loop," *IEEE Communications Magazine*, Vol. 31, No. 1, Jan. 1993, pp. 62–68.

[6]   CCITT Rec. H262, ISO/IEC 13818-2, "Generic Coding of Moving Pictures and Associated Audio," Committee Draft, Nov. 1993.

[7]   Minzer, S.E., "Broadband ISDN and Asynchronous Transfer Mode (ATM)," *IEEE Communications Magazine*, Vol. 27, No. 9, Sept. 1989, pp. 17–24.

[8]   Ohta, N., *Packet Video: Modeling and Signal Processing*, Norwood, MA: Artech House, 1994.

[9]   Piller, C., "Dreamnet," *Macworld*, Oct. 1994, pp. 96–105.

[10] Ono, S., and N. Ohta, "Super High Definition Image Communications—A Platform for Media Integration," *IEICE Trans. Commun.*, Vol. E76-B, No. 6, June 1993, pp. 509–608.

Media Integration and Image Quality

2

2.1 INTRODUCTION

Advances in electronics technologies such as low cost, large-scale memories, high-quality displays, and large-capacity storage systems are leading to a blurring of the boundaries between conventional media. For example, desktop publishing is an integration of computers and publishing. The natural evolution of image media evidently lies in the direction of media integration. In order for a general purpose image media to be able to initiate the fully fledged integration of image media, its quality must be high enough to include print, photo, and other existing nondigital media.

In this chapter, we will discuss the historical background of media integration and consider the role of the media integration platform. Quality requirements for image media integration in terms of color and spatial and temporal resolution, for example, are examined. The concept of SHD images in the context of media integration is then defined, and some sample SHD images are introduced.

2.2 MEDIA INTEGRATION

2.2.1 The Term *Media*

Broadly interpreted, the term *media* refers to all things that mediate communication between one person and another. This is made clear if we consider that *media* is the plural of the word *medium*.

This broad interpretation embraces a number of facets simultaneously, with concomitant advantages and disadvantages. The advantage is the universality of the term; the resulting ambiguity is a disadvantage. For instance, under such a broad interpretation, the single term *media* must encompass both the floppy disk, which is a physical artifact used for communication, and the mass media, an abstract structural element of modern society. If we proceed with our discussion without clarifying the meaning of the term, we invite all sorts of misunderstandings. Of course, all terms with such a universal nature possess this aspect; it is not peculiar to *media*. However, it is important to bring the reader's attention at the outset of our discussion to this characteristic of the term.

There will also be times when we use the term without completely banishing this ambiguity. For instance, the frequently encountered term *image media* sometimes refers to a physical artifact and sometimes to a social phenomenon. However, at a deeper level, it always refers to communication between people, mediated by images.

2.2.2   Media Integration

2.2.2.1   Historical Background

Media integration refers to banishing the artificial split that has occurred in media due to electronic technology. The fact that there are disparate domains for images, characters or symbols, and voice is clearly a simple artifact of technology. It all boils down to a problem of frequency bandwidth.

This is easy to understand if we consider the history of media development. In earliest human history, the advent of characters and symbols provided a secure method of recording, which was not a feature of vocal communication. Secure transmission had to wait for the invention of paper, the next major milestone. The invention of printing allowed paper media to reach large audiences over significant distances. During this period, there was no significant discrimination between image and character/symbol communications.

The split between image and character/symbol media is intimately connected to the advent of electronic media technology. The telegraph introduced real-time communication, records and telephone introduced immediate communication of vocal and sound signals, radio made possible the transmission of voice to large audiences, fax gave us transmission of drawings, and television has given us transmission of sound, voice, and moving images.

However, this division of media (or perhaps we should say of signal or information types), although essential from a technological point of view, had no clear necessity from a sociological perspective. For our readers, though, the initial technological necessity of splitting up media in this way should be obvious.

The importance of frequency bandwidth in communications was emphasized by Marshall McLuhan, the pioneering thinker of this field [1]. He classified resolutions (which correspond directly to frequency bands) into high definition, which he labeled *hot*, and low definition, which he labeled *cool*. His perception that hot media impacts the audience in a visceral way while cool media is grasped in an intellectual fashion was an important realization. This occurred, though, at a time when handling wide frequency bands remained very difficult, for both technological and economic reasons.

Optical-fiber transmission systems can handle extremely wide frequency bands, and for this reason a media integration platform for SHD imagery, which is the subject of this book, is fast approaching the realm of the possible. In McLuhan's framework, the extremely high resolution of SHD would position it as the hottest medium available, and as such it should have more of a visceral than an intellectual impact on the viewer. However, our experience with SHD has led us to the point of view that such high-resolution imagery has, in fact, a more intellectual than visceral impact, and we wish to offer a small correction to McLuhan's legacy.

Indeed, we believe that McLuhan's old observation that "hot media are less amenable to rational consideration than cool media" is in need of updating. The most salient characteristic of multimedia systems is that all kinds of image, voice, and character information are handled in an interactive fashion. Such a system facilitates an extremely high level of rational consideration, whether the immediate content is hot or cool media, and there is no question of the intrinsic amenability of the medium.

Perhaps the analysis introduced by McLuhan is similar to classical mechanics in the field of physics: it is correct under certain assumptions. However, media analysis has not yet been faced with its own version of quantum mechanics and relativity. These revolutionary changes are looming, though, in the very near future.

2.2.2.2 Media Integration and the Goal of Telecommunications

When people speak face to face, the discrimination between moving and still image media, and between character/symbol and voice media, is nonessential. In this case, people can use any media necessary for communicating with each other. Until now, however, electronic technology has perpetuated several limitations. For instance, you cannot transmit high-quality image data via telephone, nor is it possible for images rivaling the quality of print to be distributed via television. These fully understandable limitations of the technology impose constraints upon the social applications of the media. Just as the working of Japanese society is powerfully influenced by the physical fact of the existence of the city of Tokyo as the primary center, the ways in which our society now relates to media are primarily dominated by the artificial splits deriving from old technological considerations.

It should be obvious that the ultimate goal of telecommunications is to provide an environment that is as rich as actually meeting in person. Certainly, a medium that achieved this goal would not force the user to choose between image and voice or sound signals. To state the point in another way, progress must be in the direction of eliminating the limitations heretofore imposed by electronic technology. From a technological standpoint, the goal can be summarized as a requirement for transmission, switching, storage, and conversion of an extremely broad frequency bandwidth. Until now, such technology has been economically impossible, even if feasible on purely technological grounds. However, the advent of optical-fiber and ultralarge-scale integration (ULSI) technologies are bringing the goal ever closer to the realm of economic feasibility.

The goal we are discussing is, in fact, media integration. Media integration is a direct response to the society's requirements for telecommunications. Media integration will provide users with image and voice/sound at the required level of quality, and it will not force them to choose or discriminate between still and moving images. This is the point of convergence at which today's multimedia will ultimately arrive, and digital technology is the basis that makes it possible. Of course, optical-fiber and ULSI technologies are very well suited to digital technology.

We will not venture further than this into abstract considerations of media analysis, but we wish to strongly emphasize that media integration is the ultimate goal of telecommunications, and there are many social implications of this fact that have not yet been adequately considered.

2.3 QUALITY OF CONVENTIONAL IMAGE MEDIA

It is clear that a certain level of image quality is a prerequisite for media integration. In this section, we will examine the quality of conventional image media. The purpose of this section is to provide a background of image quality and so allow us to consider the requirements of media integration.

2.3.1 Color

Analyzing the human perception of color is a difficult problem. The most fundamental issues can be categorized as to whether they pertain to the color of an object or to the color of light. Also, color can be decomposed into three elements: hue, saturation, and value (brightness). However, the gray-scale colors from black to white and in between are characterized only by their value.

Whether we are dealing with the color of an object or the color of light, the color is finally determined by visible light in the range of wavelengths between 380 and 780 nm entering the human visual system. With some exceptions, color mixes subtractively when light from a light source is reflected from an object before entering the human visual system (subtractive mixture), and additively when it enters the human visual system directly (additive mixture).

In either case, the light source has a large influence on how the color is perceived. For instance, sodium lamps are commonly used on freeways because of their very high emissivity. The narrow spectrum of light that they emit, however, makes it difficult to judge the colors of objects. All of our readers are probably familiar with the difficulty of making fine color discriminations under fluorescent lighting. Human color perception is tuned to operate with natural sunlight, and this design has been built into our genes through the course of human evolution.

The characteristics used to distinguish one color from another are hue, saturation, and brightness. Hue represents the dominant color as perceived by a human observer. Saturation refers to the amount of white light mixed with a hue. Brightness means intensity. Any color can be expressed in a three-dimensional color space based on these elements. There are a number of ways of composing the three-dimensional space; Mansell space, for example, is constructed from a perceived color index. Colorimetrically measured color is expressed in the CIE XY coordinate system. The CIE is an abbreviation of Commission Internationale de l'Eclairage (the International Commission on Illumination). Readers interested in more details about color fundamentals are recommended to see [2,3].

Without exception, the signals derived from all kinds of color sensors in electronic devices are represented as red, green, and blue (RGB). These signals are generally quantized to a number of bits that depends on the application. Eight bits each for R, G, and B yields approximately 16 million colors, sufficient for most applications. Most computer systems have adopted this scheme because of its generality. However, while there is no general agreement on what the exact limitations are, human color perception is in fact far more refined than 24-bit RGB.

Cathode ray tubes (CRT) and liquid crystal displays (LCD) generate colors by combining the primary colors red, green, and blue. Printers, on the other hand, usually use cyan, magenta, yellow, and black (CMYK) as primary colors. The two systems of colors are complementary. The conversion between RGB and CMYK is performed using empirically derived parameters and is not unique. Black is usually added to the CMY because it helps to work around the tendency of black, which is produced by combining C, M, and Y, to appear brown. For printing, the optimum conversion depends on characteristics of the printing ink and requires a fair amount of empirical know-how. Furthermore, perception of color depends on solar characteristics, which differ in different regions of the globe, and on cultural and ethnic factors. A strictly objective, physical color calibration does not always produce the subjectively optimal result. Finally, mixing printing inks before printing produces colors that are subtly different from overprinting the inks.

2.3.2 Spatial Resolution

2.3.2.1 Film and Lens Performance

Film is the medium that has historically been used to record images, both still and moving. It would be nearly impossible to enumerate all of the applications in which film has found a use. Of course, the catalog provided by film manufacturers of products specialized for various applications is a good start. However, the categorization produced by this method is so coarse that it doesn't actually come close to an exhaustive listing of applications.

Film is available in a range of sizes, mainly between 35 mm and 60 mm wide, the latter also referred to as *Brownie* film. It is available in black and white or color types, and in negative and positive (reversal) varieties. Negative film is most commonly used and records colors as their complements. Positive film records the actual colors. The dynamic range of film is greater than 1,000, at least an order of magnitude greater than that of contemporary CRT displays. This contrast in dynamic range is the weakness of the CRT and the strong point of film images.

In general, fast films offer low resolution and slow films offer high resolution. Color film uses three planes, each of which detects one of three color components: cyan (C), magenta (M), and yellow (Y). The C plane is at the bottom, closest to the film substrate, while the Y plane is closest to the lens. In general, the resolution of each color plane is not the same. The resolution of color films is defined as the average value of resolution for the C, M, and Y planes.

The highest resolution 35-mm film available provides a resolution that can be calculated at about 7,000 × 7,000 pixels, but in practice it is impossible to actually achieve this resolution. To reach this resolution requires exposure with an extremely finely made laser beam scanner or a very large optical lens on a par with the ULSI exposing device. Thirty-five-mm film is usually used in such a way that a region of 24 × 35 mm is exposed, corresponding to 800 to 1,500 by 1,000 to 2,000 pixels of resolution. The actual resolution delivered depends on a number of factors in addition to the fundamental resolution of the film, such as the lighting design, the photographer's skill, and the performance of the lens used. Polaroid-style film is more real-time, but delivers only 70% of the resolution of traditional films.

The influence of the camera lens on resolution will be more apparent if we compare the 35-mm single-lens reflex cameras preferred by professionals to the 35-mm cameras with fixed or automatically adjusted focal length typically used by amateurs. Their apertures are significantly different, and if one of each style of camera is used to photograph the same scene and the resulting images printed on standard (88 × 127) photographic paper, the difference in resolution will be quite apparent to the observer. This difference will be even more pronounced if the images are enlarged and printed on a larger paper format (see Appendix 1).

Professionals whose work is used for printing, publishing, and other applications and will be significantly enlarged generally use Brownie film, which is a larger format than 35-mm. Typical Brownie formats are 60 × 60 mm, 60 × 70 mm, and 60 × 90 mm. In addition, Brownie film has no film-advance perforations. In roll films, the 65-mm movie film used by Todd AO, and 70-mm-wide films are available. As we will discuss in detail in Chapter 4, the authors use 70-mm film to capture moving-image digital data. The next step up is 4 × 5 inches, or 8 × 10 inches, but cameras that accept these sheet films are themselves rather large and awkward beasts. In particular, the 8- × 10-inch camera is essentially practical only for studio use.

If we examine not only the resolution, but also the chromatic aberration of a lens, we bring into question a number of factors that determine the performance of the lens, such as the glass material used to construct the lens, the design, and the manufacturing technologies. Ultimately, however, the only truly highly performing lenses are large ones.

2.3.2.2 Printer

Fairly high-resolution printers have, in recent years, become quite affordable, and this has been a primary factor in facilitating the advance of desktop publishing. Experience shows that a resolution of at least 300 dots per inch (dpi) is required for printing text. The two printing technologies that satisfy this requirement and are commercially available are laser beam printers and ink jet printers.

With slight differences between models, ink jet printers basically squirt extremely small droplets of ink. These printers consume little power and are compact, quiet, and inexpensive. They typically deliver resolutions of 300 to 720 dpi, quite adequate for printing text. That is, even at a 10-point size, the difference between fonts is readily apparent, and jaggies are essentially invisible. The drawback of ink jet printing is that the quality of the paper has a large effect on the quality of the output. The other drawback with ink jet printers is that they can only reproduce very small color areas in the CIE chromaticity diagram, because they use only four ink colors.

Laser printers use the same basic mechanism as electrostatic copying machines. The power required to fix the toner to the paper makes these printers bigger power hogs than ink jet printers. Several years ago, laser printers were the only economical alternative that delivered 300 to 400 dpi, and as such were an important factor in making desktop publishing possible. Currently, laser printers deliver higher resolution, 400 to 1,500 dpi, and higher speed than ink jet printers. Laser printers delivering 1,000 dpi and higher are still quite expensive and are thus limited to professional use.

The difference between 300- to 400-dpi and higher resolution is much more apparent when dealing with photos and other prints than it is with text. When photos and other images with gray scale content are displayed on a black and white output device such as a printer, dithering must be employed, and dithering at a resolution of 300 to 400 dpi degrades the image considerably.

The highest quality color printers available are based on silver salt film, followed by dye sublimation and ink jet printers. The color balance of dye sublimation and ink jet printers approaches that of printing, but in order to produce the very subtle colors possible with printing ink, a special-purpose color balance is generally necessary. The resolution of silver salt film is 300 to 400 dpi, and the resolution of the other technologies is about 50% of that figure.

2.3.2.3 Printing

Printing is the technology that has supported the mass media ever since it was invented by Gutenberg. There are various types of printing, with typesetting being the principal technology in current use. The movable type invented by Gutenberg remains in use for only a few specialized applications.

Printing can be categorized according to whether it relies on relief (concave), intaglio (convex), or lithography (plane). The term *lithography* is essentially synonymous with *offset printing*. Strictly speaking, the terms are different, with lithography referring to the printing plate and offset printing referring to the mechanism employed. But in modern usage, lithographs are used almost exclusively for offset printing, so we take the liberty of calling them synonymous. Intaglio is also called *gravure printing*, and is used in very-high-quality applications, such as art printing. It is well suited to processes that use a wider variety of inks than commercial printing does.

The resolution of printing processes is very high, corresponding to 2,400 dpi. However, printing resolution is generally specified in lines per inch (lpi), and this figure takes into consideration the angle of the lines. The conversion is usually calculated at 16 dpi per 1 lpi, which yields a figure of 2,400 dpi for the typical printing resolution of 150 lpi. However, this is only one way in which dpi and lpi might be compared and is used only as a rough figure representing image quality when, for instance, preparing original materials for printing. In fact, dpi and lpi are not directly related, and it is not the case that printed matter produced in large volumes actually delivers these resolutions. The quality of the paper used has a very significant effect on the ultimate resolution, and printing onto newsprint results in a large degradation. In fact, the output of laser or ink jet printers with resolutions above 300 dpi is difficult to distinguish with the naked eye from printed materials or text. These levels of resolution are critical for achieving esthetically acceptable layouts and fonts.

2.3.2.4 CRT and LCD Displays

The CRT is a vacuum tube that forms an image by striking phosphor with an electron beam, causing it to emit light. The human eye cannot resolve point light sources separated by less than about 0.2 mm; they are interpreted as a single source. Therefore, the resolution of the CRT saturates at about 2.54/0.2, or 120 dpi. Leaving some room for a margin, the most frequently used resolution is 72 dpi (for each of the R, G, and B color pixels).

The intensity of the CRT's electron beam is varied to produce varying intensities of light or color. The intensity of the beam must be increased in order to generate higher intensity light, and this results in degradation of the resolution. The negatively charged electrons in the beam repel each other, causing the beam to spread.

The emissive characteristics of the CRT are dominated by the emissive characteristics of the phosphor used, affecting the color reproducibility and the speed with which the image can be changed. The emission from the phosphor decays gradually when the electron beam stimulation ceases.

LCDs, like CRTs, use point light sources to form an image. The light sources are not phosphors, but white light which is passes through a set of RGB filters. The adjustment of light intensity is performed by the liquid crystals after the light has passed through a filter. With currently available products, this adjustment is coarser than that of CRTs, but advances expected in the near future should place LCDs on a par with CRTs in this regard.

The color reproducibility characteristics of LCDs are broader than those of CRTs, because the spectral range of the combination of color filter and light source used in the LCD is broader than that of the phosphors used in CRTs. In addition, the point sources produced by an LCD display do not shift position on the screen as those of the CRT do. This shifting is caused by the CRT's use of a magnetic field to steer the electron beam.

Both CRTs and LCDs are available in direct-view and projection models. Projection models are useful for projecting onto large screens. However, projection introduces a number of weaknesses: the image is not as bright as with direct view and the image is affected by the characteristics of the screen.

The highest resolution CRT currently available is a 28-inch model with $2,000 \times 2,000$ pixels. CRTs of this class are available from a number of manufacturers. The highest resolution LCD currently available is a 12-inch model with 640×480 pixels. However, as LCD technology advances to provide large-format models at low cost, the advantage of the LCD display in terms of small depth and low power consumption will allow it to replace the CRT. This displacement is expected to begin with applications that demand very high precision.

The dynamic range of the LCD is not yet on a par with that of the CRT, but this should be corrected within a few years. It should be possible to increase the pixel count of the LCD without much difficulty. The LCD can be adapted to broad-bandwidth input signals by providing it with multiple input ports. The LCD, unlike the CRT, is not influenced by the earth's magnetic field when it is scaled to large formats at high resolution. Thus, scaling up the size is the only significant technological issue remaining. However, LCD display sizes have been growing at a pace of about 1 inch per year, and this pace is expected to continue. Before long, the LCD will replace the CRT as the predominant display technology.

2.3.3 Temporal Resolution

2.3.3.1 The Smoothness of Motion

Moving images are displayed as a series of still images, called *frames*. The perceived smoothness of the motion depends on how many frames per second are displayed. If we compare recent animation on television with old-fashioned Disney animation, the influence of their higher frame rates on smoothness is apparent.

Increasing the frame rate causes the motion to be smoother, but the cost is an increase in the volume of information. Because the information required to represent an image is quite large to begin with, the simplest way to decrease this high data rate is to decrease the frame rate, and this simple method is frequently seized upon. Of course, decreasing the frame rate detracts from the smoothness of the displayed motion. The H.261 standard for coding moving images, principally used for video telephone, uses frame rate reduction despite this loss of smoothness. It operates at about 10 frames per second, but the rate varies during operation. The lack of smoothness in the moving images is entirely apparent to even a casual observer.

Let us look at the frame rates of conventional moving-image media. Cinematic film uses a frame rate of 24 frames per second, but each frame is displayed twice by the shutter. This is essentially a method of increasing the smoothness of motion and uses the fact that film is displayed in a very dark environment and projected onto a very large screen. The NTSC television standard used in the United States, Canada, and Japan operates at 30 frames per second. The PAL and SECAM standards mostly used in Europe operate at 25 frames per second. (PAL and SECAM use a lower frame rate than NTSC, but scan 625 lines, providing higher spatial resolution than NTSC's 525 lines.) Both of these standards, though, are interlaced.

Interlace has been used for a long time to reduce TV signal bandwidth while maintaining its temporal resolution. Interlaced systems display the odd-numbered lines in one scan of the screen and the even-numbered lines on the next scan. This involves a trade-off between spatial and temporal resolution. Each of these scans is called a *field*, and two fields make up a frame. This means that NTSC displays 60 fields per second. This style of interlace is called *2:1 interlace*.

Interlace is fairly effective for improving the quality of natural types of images. However, there are types of images and conditions for which it is not effective. In particular, still images and stop frames tend to flicker quite noticeably. The effectiveness of interlace is based on the observation that human visual acuity is reduced for moving objects. However, it has been pointed out that the reduction in visual acuity does not always hold, and we must keep this fact in mind [4,5].

Another consideration with film is that 24 frames per second is sometimes insufficient to display smoothly the motion of very quickly moving objects. The Showscan format uses 70-mm film at 60 frames per second to counter this problem. However, this format has not become widely used because of problems such as the increased cost of film stock.

2.3.3.2 Flicker and Display Devices

Flicker is another factor that decreases the quality of images. Flicker refers to the effect that makes images blink due to low frame rates. In fact, flicker is much more annoying than reduced spatial resolution. If the reader will consider that computer interfaces use blinking, which is basically the same phenomenon as flicker, to draw the user's attention to, for instance, the cursor, this fact will be obvious. Thus, even for a system with adequate spatial resolution, if a viewer notices flicker, the subjective quality assessment of the image will be greatly reduced.

Staring for a long time at a flickering screen is a sure way to tire yourself. For this reason, computer monitors are designed to operate at frame rates of at least 50 frames per second, so that flicker is not noticeable. Both interlaced systems (at least 100 fields per second) and noninterlaced systems are used. There is little benefit when the noninterlaced frame rate is increased above 100 frames per second, so computer monitors generally use a noninterlaced scan format.

The way in which flicker is perceived depends on the image signal strength and the display device characteristics and brightness. Because flicker is essentially a type of aliasing, it does not appear when the image signal does not contain high-frequency spatial components. However, images without high-frequency spatial components lack sharpness. For images of the natural world, this lack of sharpness is not usually a problem, but for most other types of imagery, the image quality will be perceived as poor.

The phenomenon of flicker is perceived entirely differently for CRT than for LCD displays. Flicker is less apparent on LCD displays due to the slower decay characteristics of light emission from LCD devices. Because we expect that their large-format advantages and other reasons will cause the LCD to displace the CRT as the predominant display technology, it is quite likely that a noninterlaced signal of about 60 frames per second will be adequate to provide flicker-free images.

2.4 SHD IMAGES FOR MEDIA INTEGRATION PLATFORM

As defined in Chapter 1, SHD imaging refers to high-quality digital imaging for media integration. In this section, we will consider some requirements of SHD imaging for media integration, focusing on image quality, aspect ratio, display systems, multiple-window concepts, and relationship to computers [6].

2.4.1 Image Quality

There are still many aspects of the coming media integration that are unclear. What is quite clear, however, is that media integration will place imagery of all types on a single display system. If this is the goal, how many pixels will be required? How about the frame rate to display the images? That is, what level of image quality is called for? There is no single answer to this question. Or perhaps we should say that there is an infinite number of answers. The answer depends on the type of imagery and its application, and there are virtually countless combinations. With this caveat in mind, we will attempt to formulate some sort of common denominator.

2.4.1.1 Spatial Resolution

It is very informative to examine the sizes of film and the corresponding image quality to answer the above-mentioned question in terms of spatial resolution. Film presents exactly the same problem of determining the optimal resolution, which depends on the scene to be photographed and on the application in which the photograph will find a use. Currently, standard sizes of film for stills are 35-mm and 60-mm (Brownie) for roll film, and 4- × 5- and 8- × 10-inch sheet films. As we discussed earlier, we will discuss still film in two categories, 35-mm and the larger formats (Brownie and sheet films). The former category can be considered for amateur use and the latter for professional use. Certainly this is not a hard and fast rule, and there are many exceptions. This division is of interest to us in our examination of media integration, because it demonstrates the threshold resolution value below which the medium is of limited application for professional use. Let us take a look at the resolutions involved.

Empirically, we know that the threshold lies somewhere around 2,000 × 2,000 pixels. The authors have carried out some experiments that validate this figure. Our discussions with professional photographers and printers also corroborate this result. Their common opinion is that it is almost impossible to achieve a resolution of 2,000 × 2,000 with a 35-mm camera and film. The reason why we focus our attention on the opinions of professionals is that we are trying to analyze what the technical requirements will be for media integration, and the requirements of the professionals, who are the image suppliers, are easier to pin down than the requirements of the viewers, who are the consumers.

At the risk of sounding redundant, let us emphasize again that this division between professional and amateur use is a very rough one, and that there are many exceptions to the rule. Furthermore, there is no hard and fast reason why film must be divided in this way; a certain amount of historical accident is involved. Finally, film technology continues to advance; improvements in resolution are particularly noteworthy.

In the end, though, we are left with the undeniable fact that the resolution of 2,000 × 2,000 represents some important threshold. It is possible that this threshold has been brought into being by the printing industry. In any case, this is a very reasonable figure to use as the first achievable target for a media integration platform, although we still have to cover higher resolution for some special applications.

2.4.1.2 Bit per Pixel (Accuracy of Pixel Quantization)

The number of bits required to digitize a pixel is a function of the application. For example, eight bits each for CMYK is adequate in printing applications. On the other hand, if we want to achieve the same quality of film, 10 bits will be required for each color because it is said that the dynamic range of film is greater than 1,000. However, there are many applications, including medical diagnosis, for which 8-bit accuracy is enough. It seems to be reasonable to settle upon an 8-bit threshold for pixel accuracy per color for color images. This is based on our anticipation that almost all intellectual activities are adequately supported by 2,000 × 2,000 images displayed with 8-bit accuracy. Of course, it is desired that the original images be digitized with 10-bit accuracy for preservation.

2.4.1.3 Temporal Resolution

Another important value for evaluating the quality of imagery is the temporal resolution. What level of temporal resolution will be required for media integration? The answer to this question is clearer than for spatial resolution. Research and development of computer displays, as well as their widespread acceptance, gives us the answer. A temporal resolution of around 60 frames per second is needed, about twice the rate provided by current television. At this rate, flicker is not perceptible, no matter what type of image is displayed. The perception of flicker is complex, depending on the environment in which the viewing occurs, and how the viewer looks at the screen. A rate of 70 frames per second essentially eliminates flicker, regardless of the viewing conditions, while 60 frames per second leaves very little margin. However, 60 frames per second should be adequate for LCD displays, which are expected to predominate in the near future, as discussed earlier.

2.4.1.4 Scanning Method

As already discussed, interlace refers to the technique in which the signal transmits the even lines in one pass and the odd lines in the next pass. Interlace is effective for increasing the smoothness of motion, but only when the image does not contain significant high-frequency components. It is utterly without benefit when the image signal contains high-frequency components. Because images of natural scenery tend not to contain significant high-frequency components, there are benefits to the use of interlace under certain circumstances.

However, media integration must be able to handle signals with high-frequency components, so interlace provides no benefit at all. Therefore, a strictly noninterlaced system or a progressive scanning system, which scans all the lines of the display in sequential order, must be adopted. Recently, extended television (EDTV), which enhances television quality while keeping its basic format, and other systems have also adopted noninterlaced scanning in order to improve image quality [7]. The spectral characteristics of an interlaced signal are far more complex than those of the original, noninterlaced signal. This is an undesirable feature from the point of view of efficiently coding the video signal.

2.4.1.5 Summary

In Figure 2.1, we have plotted the spatial and temporal resolutions of traditional media in two dimensions. The position in this analysis of SHD imaging with media integration as its goal, should be apparent. The distinctions between still and moving imagery should also be apparent. The authors have enlisted a number of professional photographers, printing experts, medical experts, and experts from the optics industry. With their assistance, we investigated the hypothesis that 2,000 × 2,000 is an important spatial resolution threshold by displaying 2,000 × 2,000 images digitized at eight bits per color (R,G, and B) at the frame rate of 60 with a progressive scanning method. Not one of our respondents opined that the pixel count was too low, the resolution inadequate, or the picture quality insufficient.

2.4.2 Multiple-Window Systems

The term *media integration* has two meanings. It refers not only to the ability to display multiple media simultaneously, but also to the ability to display multiple types of image data simultaneously on a single screen. This requires a particular display style. The most natural style is the windowing system, which allows images to overlap. This is essentially a metaphor for a desktop, pioneered by the Xerox Alto machine, and has been adopted by today's PC.

Allowing the windows to overlap means that a smaller display can suffice. For the foreseeable future, display technology will continue to impose limits on maximum screen size. Therefore, the current style of overlapped windows should predominate.

Of course, one could use multiple display devices, but this would be prohibitively expensive. In addition, when we consider the problems involved in moving images about between the displays and the like, the multiple display solution loses a great deal of its initial attraction. Also, if an overlapping window system provides a very usable interface for changing the precedence of

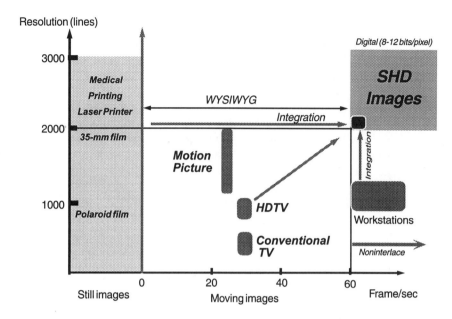

Figure 2.1 Characteristics of SHD images.

windows, it is not really cumbersome for the user. In fact, most overlapping window systems provide an interface for manipulating windows that is exceedingly easy to use. The limit on the number of images that can be viewed comes not from the windowing system, but from the type of display used. We call it a *desktop metaphor*, but in fact there are many times when it is easier to use than an actual physical desktop!

Figure 2.2 (see color insert) illustrates an example of such a window system in operation. It is clear that a window system similar to this one will be used in any media integration system. Whereas current window systems display multiple windows, each of which tends to be filled with text, the windows of the media integration system will be just as likely to contain high-resolution moving or still images and graphics. The commercial product QuickTime is an approximation of the media integration system of the future, only at lower resolution.

Multiple-window systems have finally realized the Memex system proposed by Vannevar Bush in 1945 [8], which may be considered the seed of the present-day PC. Bush imagined Memex as a desktop system that would provide easy, instant search and access to files, or even an entire library, allowing the results to be processed or used on the spot. Because the overlapping order of images on the display can be easily changed, the whole paradigm is actually considerably more functional than the mountains of books and documents that tend to occupy a real-life desktop. For this reason, the multiple-window user interface should be advanced further and become very widely used. The importance of this fact for media integration cannot be overstated.

2.4.3 Aspect Ratio

The aspect ratio of an image is the ratio of its height to its width. Present day television has an aspect ratio of 3:4, while HDTV's aspect ratio is 9:16. The debate over the "correct" aspect ratio for media integration shows no sign of converging to a single answer. There is no single aspect ratio that is optimal for all kinds of images and documents.

The reason why something closer to agreement on a single aspect ratio has come about in the TV/HDTV wars is that the goal of HDTV is to display cinematic films. However, even the aspect ratio of films is not uniform, although most are distributed around 5:9.

The furor over the aspect ratio of HDTV seems to be dying down for two reasons. One is the widespread acceptance of multiple-window systems in the PC world. The other is the market penetration of wide-screen TV. With each of these systems, users can select from a number of display modes. Thus, there is less resistance to systems where users select the display mode according to their needs.

The Macintosh display shown in Figure 2.3 can be turned sideways, allowing users to select either of two aspect ratios. Users select the aspect ratio that best suits the imagery they are working with. The trend toward display flexibility that this product represents will continue, and users will be able to choose the display mode that best suits the viewing conditions and the imagery being viewed.

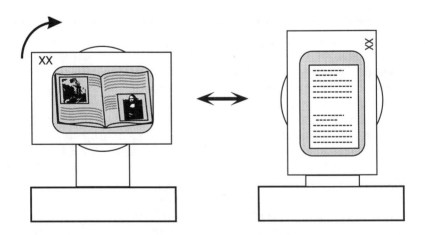

Figure 2.3 Two-way usage of a Radius's "Color Pivot" display.

The most important thing is to be able to freely select the display mode that suits your needs, and digital image processing technology makes this almost entirely achievable today. So the great debate over the "right" aspect ratio should die down into quiet discussions of guidelines rather than a universal specification.

2.4.4 All-Digital System

The ultimate media integration platform or multimedia platform should subsume all conventional image media. Conventional image media have been essentially based on analog formats. Recently, however, digital technologies have been introduced to some imaging systems. For instance, digitized images are usually used in the advanced printing industry. For editing TV signals, digital technologies are also employed [9].

There is no doubt that the multimedia platform should be all-digital. The term *all-digital* means that not only is the system itself (including the analog-to-digital (A/D) and digital-to-analog (D/A) converters) entirely digital, but the processing that occurs is also clearly digital. Concrete examples are the operations of adaptive quantization, vector quantization, and entropy coding (Huffman coding and arithmetic coding) (see Chapter 3).

All-digital systems can provide the following advantages: high stability, high accuracy, and high flexibility. Flexibility is indispensable for the future multimedia platform systems, because there will be a variety of demands imposed by as yet unknown applications. Although nonprogrammable systems deliver higher processing performance for fixed functionality, programmable digital systems are definitely advantageous when flexibility is required (see Chapter 6).

2.4.5 Relationship to Computers

Since computer interfaces have come to use icons and the like, the display of processing results has significantly improved. That is, when a computed result is to be simply displayed as a character string, all you need to do is to convert the internal data format (floating or fixed point) into ASCII character codes. Of course, there are a fair number of nonprinting character codes. These differ in function from the letters (e.g., A, B, C), digits (e.g., 0, 1, 2), and symbols (e.g., !, @, #, $). The nonprinting character codes serve, for instance, to specify the position of the printable characters.

If characters are all that need be displayed, it is not very difficult to specify their location on the screen. You simply distinguish a certain location as the origin and specify locations in units of characters to the right or left, above or below, that location. This system can be implemented by interspersing nonprinting codes (typically escape codes) with the printable characters to specify their locations. Screen scroll, too, can be easily implemented in this way. Some nonprinting code (typically the line feed code) is used to specify scrolling.

However, simple character strings are not sufficient when the screen must display image data. It is necessary to be able to specify any pixel location on the screen, as well as the data (bit pattern) to be displayed there. That is, if we take 0 or 1 as an element, the contents of a rectilinear three-dimensional region must be specified during each frame interval. This region has sides equal to the horizontal and vertical pixel count of the display and depth equal to the number of bits per pixel. This results in an enormous amount of data. Indeed, for a 2,048 × 2,048 SHD image, it comes to about 100 Mb. But it is frequently the case that the actual image data generating the screen is far smaller than this. Therefore, it is necessary to expand the 0s or 1s necessary for display from the image data into a rectilinear three-dimensional region within a frame interval. In the same way, taking 0 or 1 as elements, a rectilinear three-dimensional region must be converted in an appropriate way for the next screen during a frame interval.

Page description languages, such as PostScript, are standard languages that are used to carry out this expansion. In a certain sense, this can be considered an extension of the idea of nonprinting codes. Coordinate transformations are commonly used. Affine transforms are the most common transforms involved [10].

These calculations must be reliably completed, on huge data sets, within a very short time, so very-high-speed processing is necessary. The color lookup table (LUT) and graphics processor (GP) are aids to speed the processing. A LUT is a table that contains the result of a calculation as an RGB bit pattern, and is a combinatorial circuit. The graphics processor is a special-purpose processor which can calculate certain patterns extremely rapidly. Of course, this assumes that the structure can be represented as a rectilinear three-dimensional region with elements that are 0 or 1, and that vector processing that uses a pipelined repetition of routine processing is possible.

Today's PCs, whether large or small, all contain such a system and can handle all sorts of images with great facility. This is the most salient characteristic distinguishing them from the computers of yesteryear. The SHD imaging system is a natural extension of this type of system. It is clear that PCs and SHD will merge into a single class of computer. However, the term *personal computer* may not remain with us indefinitely.

2.5 SHD TEST IMAGES FOR REFERENCE

2.5.1 Overview

There are a number of significant differences between text and images. First, text comes in a variety of fonts, whereas images do not. But for a variety of reasons there are many times when test images are required in image applications

The most familiar example is probably the color bar test pattern used in television. This geometric pattern is used mainly to evaluate resolution and color reproducibility. Conventional television is a low-resolution medium, with limited ability to reproduce colors, and as such it can be adjusted using a simple geometric pattern with only ten colors; that is, an objective measurement correlates well with the subjective evaluation of the result. For this reason, the color bar test pattern is widely used.

Such test patterns are also required for SHD imagery, especially for evaluating coding quality. However, geometric patterns, such as the familiar color bar pattern, are essentially useless in conjunction with SHD imagery. As with high-quality printing, these sorts of objective tests are useful to verify that the system is performing at some minimum level, but they are not useful for evaluating the actual image quality.

2.5.1.1 Difficulty of Color Tuning

The color checker pattern with 300 or so colors can be used to accurately evaluate the color reproduction of a system, but, again, this is useful only as a general guideline or to verify that the system achieves a certain minimum level of performance. It is not used to evaluate quality. The reason for this lies in the fact that the color checker pattern is tuned for use in evaluating color reproducibility. Resolution can be evaluated as a single figure, but the same is not true for color reproduction. Color tuning is an issue of balancing R, G, and B, and tuning one color will strongly influence the reproduction of other colors. Of course, tuning any color amounts to modifying R, G, and B all at once, so this result should not be surprising. What it all means is that color tuning is a vexatious problem; sometimes it makes you want to scream!

In fact, if color tuning a conventional TV with a color bar containing only 10 colors is so problematic, we can well imagine that color tuning that must bring into line at least ten times that many colors may well be beyond human skill or patience to perform. The tuning can in fact be performed with the aid of an optical colorimeter. While it still requires some time to perform, the task is doable.

However, even if we decide that we will tune and optimize the color reproduction with an optical colorimeter, our job is still not finished. The next hurdle we face is that there is no widespread agreement on just what an optimum color tuning is. Our own research has shown that tuning can certainly be accomplished to some extent, as shown in Chapter 7 in more detail. However, it is quite obvious that, when the tuning is complete, the colors are still off to such an extent that anyone who observes the screen can tell. So, while this sort of procedure is not without meaning, it certainly does not provide comprehensive color tuning.

Perhaps the differences that arise in the optimal values are due to the limited number of colors used in the tuning. In that case, how many colors would be sufficient? The reader should keep in mind that even with only 30 colors to adjust, the procedure requires several hours to complete.

Of course, all of this is by no means new information. The tuning method that is most widely used, therefore, is to make a very rough adjustment to each of the colors on the color checker, and then to use several natural scenes to check the color reproduction.

2.5.1.2 Motivation for Creating Test Images

The basis of using test images is that if the image quality is high enough, human perception is adequate for color checking. How close does the image appear to the real thing? Of course, there is no particular theory that supports this procedure; it is simply empirically derived. However, it is quite effective, and for that reason is widely used in the printing industry.

The main problem with this procedure is that it is not clear what sort of images, and how many, should be used. Of course, since the entire procedure is empirical, experience can serve as a guide. However, this is merely a seat-of-the-pants type of guess, and we have not yet addressed the issue of how much time is required. And let us also emphasize that for images like these, the artistic value is also very important.

This section describes some of the intermediate results gained when the authors created SHD test images. A suite of SHD test images is being assembled using the digital standard color image data (SCID) as a reference [11]. These images are available to other investigators who wish to evaluate SHD images.

2.5.2 Creating the Test Images: Photography and Digitization

As we mentioned earlier, the resolution of 35-mm film never exceeds 2,000 × 2,000, so we use Brownie film and 4- × 5-inch film to create our test images. Of course, if there were a video camera or an electronic still camera of the requisite resolution, we would use it, but at the current time, film is still the best solution. This is discussed in detail in Chapter 4.

The studio photographs used a 6,500K strobe, with positive films manufactured by Kodak and Fuji. The cameras were manufactured by Hasselblad and Mamiya, and their sizes were 6 × 6 cm and 6 × 7 cm, respectively. Siner's 4- × 5-inch camera was also used.

After being developed, the films were scanned and converted to digital data. The scanner uses a linear charged-coupled device (CCD) sensor manufactured by Scitex, uses a halogen lamp, and produces 16 bits for each of the R, G, and B channels at the output of the A/D converter. Each datum was rounded to 8 bits using a nonlinear table embedded in the system. Of course, the output data contain all sorts of noise, so there is no assurance that 16 significant bits of data are produced. (The system is described in Chapter 7 in more detail.)

One of our test images is reproduced in Figure 2.4 (see color insert), including the color checker and some other images. The original photograph was taken by Issei Isshiki.

2.5.3 Storage

Film is composed of the film base and the film emulsion, each of which is a chemical product and susceptible to change with time. The film base is the more stable of the two, but nonetheless may undergo significant changes with time, depending on the environment in which it is stored. The stability of the emulsion is affected by the color coupler, whether contained in the emulsion or in the developer. The coupler contained in the developer is the most stable of these, but even so it is well known that it, too, can change significantly with the passage of time. The lifetime of the color in a typical 35-mm film is less than 20 years due to impurities in the developer and due to the environment in which the film is stored.

For this reason, the long-term stability of glass photographic plates is particularly noteworthy. This advantage is offset by the fragility, weight, size, and thickness of glass, which render it burdensome to handle. Further, it is entirely unsuited to applications with moving images.

Although film cannot be rendered absolutely stable, sufficient care in developing and storing it can extend its life significantly. For this reason, we use a film storage cabinet that provides storage environments for film materials. The cabinet's environment is 10° C lower than room temperature and its humidity is 30%. The film used to create our test images was stored in this cabinet from the time it was developed until it was digitized.

Digitizing an image solves the problem of film instability. A digital representation of an image does not change with time and has a virtually infinite lifetime. If the data format of the digital representation includes error-correcting coding, the original data can always be recovered.

Of course, plastic is generally used as a substrate for storage of digital data. This is true of magnetic tape and of CD-ROM. These plastic products, like film, are chemical products and will change with the passage of years. However, more metal is used than is the case with film, so they may be a bit more physically stable than film. Nevertheless, in either case, they are subject to deterioration over time.

However, with digital data, when the physical medium has deteriorated over time, the original data may still be perfectly recovered and rerecorded onto a new physical medium. This is what truly distinguishes digital storage from all previous media. So, by transferring the stored digital data onto a new physical media at some appropriate interval of time, absolutely lossless storage can be ensured. Of course, the correct interval of time will depend on the characteristics of the physical medium used and the effectiveness of the error correcting code. What we wish to stress is that the capability of maintaining information, whether text or images, indefinitely without change is an absolutely new phenomenon and is certain to have an important impact.

2.6 SUMMARY

We have discussed the technological issues surrounding media integration and have described the concept of SHD images as the platform for media integration. The technological requirements for SHD images as the platform were examined, focusing on image quality (e.g., resolution, scanning method), aspect ratio, and other related matters. In order to include conventional image media, SHD images should have a resolution of more than 2,000 × 2,000 and a frame rate of more than 50 with progressive scan. The necessity of SHD test images was discussed, and the authors' effort to create SHD test images was introduced. The images are now being distributed for research purposes. Although they are still intermediate results, the authors have enough input from printers, cinematic film producers, cameramen, stylists, and makeup artists to be certain that the test images are useful.

The meaning of the term *media* may yet undergo significant changes. Furthermore, there is no clean, crisp image that we can point to and say "this is what we mean by media integration." The authors believe that the very reason for this vagueness is the richness of possibilities that are unfolding. It is expected that SHD imaging, the subject of this book, will contribute to changing our concept of media and there is no room for doubt as to the usefulness SHD imagery as the media integration platform.

References

[1] McLuhan, M., *Understanding Media*, New York: McGraw-Hill, 1964.
[2] Hunt, R.W G., *The Reproduction of Colour*, 4th ed., Tolworth, England: Fountain Press, 1987.
[3] Gonzalez, R., and P. Wintz, Chap. 4.7 of *Digital Image Processing*, 2nd ed., Reading, MA: Addison-Wesley, 1987.
[4] Miyahara, M., "Analysis of Perception of Motion in Television Signals and Its Application to Bandwidth Compression," *NHK Technical Journal*, Vol. 27, No. 4, 1975, pp. 141–171.
[5] Watanabe, A., T. Mori, S. Nagata, and K. Hiwatashi, "Spatial Sine-Wave Responses of a Human Visual System," *Vision Research*, Vol. 8, No. 9, Sept. 1968, pp. 1245–1263.
[6] Ono, S., and N. Ohta, "Super High Definition Image Communications—A Platform for Media Integration," *IEICE Trans. Commun.*, Vol. E76-B, No. 6, June 1993, pp. 599–608.
[7] Schäfer, R., "Recent Advances in Digital Transmission of HDTV," *Proc. Conf. on High Definition Video*, SPIE, Vol. 1976, April 1993, pp. 2–11.
[8] Bush, V., "As We May Think," *Atlantic Monthly*, Vol. 176, No. 1, July, 1945.
[9] Sandbank, C.P., *Digital Television*, Chichester, England: John Wiley & Sons, 1990.

[10] Adobe Systems, Inc., *PostScript Language Reference Manual*, 2nd ed., Reading, MA: Addison-Wesley, 1990.

[11] Japanese Standards Association, "Standard Color Image Data," 1993.

Image-Coding Algorithms **3**

3.1 INTRODUCTION

To represent images digitally generally requires vast amounts of binary data, which means that bit rate reduction is extremely important for efficient storage and transmission. Bit rate reduction is a means of encoding and decoding digital signals so that the original signals can be compressed with no or tolerable loss of information. The technologies for image signals are usually referred to as image compression technologies. Image compression can be thought of as using the structure of the image data stream to reduce the original data to a more compact form.

In this chapter, we will focus on fundamental algorithms for image compression and present an overview of the standardized algorithms like JPEG and MPEG. The results of applying those standard algorithms to SHD images are also shown. Simulation results will show how much bit rate will be necessary for real-time SHD moving-image transmission.

3.2 FUNDAMENTAL ALGORITHMS

3.2.1 Overview

Bit rate reduction may be characterized by whether it is reversible or irreversible. With reversible coding, it is possible to regenerate the original digital data completely. In irreversible coding, the decoded data merely resemble the original data, with a great deal of variation in the degree of resemblance from one system to another.

Reversible coding, also called *lossless coding*, can be used for compressing several types of digital data files and often employs bit arithmetic. Examples of reversible coding are the Lempel-Ziv algorithm, commonly used for file compression, and the entropy coding method developed from information theory [1].

Both are variable-length coding (VLC) methods which operate on variable units of processing. Neither method is restricted to image data, and both are used in a wide range of applications, including the compression of audio signals.

Irreversible coding methods, also called *lossy* methods, attempt to approximate the original digital data, usually by exploiting the characteristics of human perception and various image data characteristics. These methods often use arithmetic operations with finite word length and are thus unavoidably subject to a certain amount of computational error.

Lossy methods often capture the image data structure in terms of frequencies or corresponding domains that represent characteristics of the data (i.e., feature extraction). The basic scheme to transform the original signal into these domains is called *transform coding*. Another method of frequency analysis is to prepare a set of filters with different pass bandwidths (called a *filter bank*) and obtain various signal values at the output port of each filter. The main issue in this method, usually called *subband coding*, is to determine the characteristics and number of filters used.

Obviously, these two methods are not completely independent. The nature of their relationship has been much discussed, and recently a third transform method, called a *wavelet transform*, has been discovered, which is intermediate between the two. Although no reports have established that a wavelet transform offers superior high-quality image coding, there have been reports of favorable properties; for example, there is no concentration of errors at the edges of images encoded at low resolution.

The compression ratio of lossy and lossless coding methods may be stated simply as follows. Lossless methods have a maximum compression ratio of about 2:1; anything better than this is thought to be impossible, with the exception of image data having a special structure. The ratio of lossy methods depends on the outcome of the ongoing discussion about what constitutes "permissible distortion." Obviously, this value will vary greatly according to the type of image and will also depend on how we define the concept of "distortion." Given these caveats, we can analyze the compression ratio of lossy coding as follows. The distortion introduced by image compression rates up to 10:1 is almost undetectable to humans, including professionals. The compression rate of 20:1 is undetectable to most amateurs. Any compression beyond this point is a question of how much distortion is to be tolerated, which depends on how the image is to be used.

International standards, both lossy and lossless, are rapidly being developed for image coding, the most active groups being the JPEG for still images and the MPEG for motion pictures. Note that JPEG standards are also used for motion pictures, a subject that will be discussed in more detail below. MPEG uses interframe correlation in order to increase the efficiency of coding motion pictures, but this makes the MPEG method troublesome to use with still images, because it requires data from several frames before and after the frame in question. Not only is the MPEG method difficult to use with still images, but it also decreases the coding efficiency of editing devices because of their frequent starts and stops. Therefore, the JPEG method is used for motion picture editing systems, even though its coding efficiency is lower on a per-frame basis.

3.2.2 Transform Coding

The most basic compression technique is to reduce the bit rate by taking advantage of any correlation in the signal. If we consider the frequency distribution of image signals containing strong correlations, the signal power must be concentrated in the low-frequency region. The basic idea behind transform coding is to transform the original signal into the frequency domain to use this concentration for compression. In general, it is possible to exploit for compression any systematic bias in components of the signal so that we can remove redundancy more efficiently.

Generally the transformation is accomplished by a discrete Fourier transform (DFT); in particular, the discrete cosine transform (DCT) is preferred because of its superior extraction of low-frequency components. From a theoretical point of view, the optimal transform is the Karhunen-Loeve (K-L) transform, but there are many problems obstructing its practical application. While DCT is similar to the K-L transform in many respects, it has fewer barriers to practical implementation [2].

In mathematical notation, transform coding is a method for coding image data A, which is a set of two-dimensional data, into the two-dimensional data set B, and then generating code from the elements of data set B. In other words, $B = FA$, where A, B, and F are each $n \times n$ matrices. Matrix A is obtained by allocating a certain-size block to the original image, and matrix F is applied to image data A in order to calculate matrix B. Normally, the value of n will range from 8 to 64, but it may range as high as 1,000 to 4,000. Unless some special consideration applies, n is normally a power of 2, which is most suitable for high-speed algorithms. Various transforms are possible, depending on the matrix used for F.

The efficiency with which B can be calculated depends on the structure of F. Matrix A is normally written as n rows and n columns, and the calculation is repeatedly applied to the rows and columns. Although this method has been established as a standard, it is not always the fastest way.

One problem common to these transform coding methods is that clearly discernible block distortions readily appear at the borders between the blocks, which are used to increase efficiency. Block distortion can be reduced by using overlapping blocks to obtain matrix A from the original image, but this redundancy increases the number of calculations which must be performed.

3.2.3 Subband Coding

Subband coding refers to the practice of using filters with different pass regions in order to apportion the image signals into various areas, and then using the signal values output by each filter port in order to generate the encoded data. The basic idea of subband coding is to take advantage of bias in the frequency spectrum of the image signals by dividing the signal into multiple bands. The spectral distribution of ordinary image data is biased toward the lower frequency regions. If data from this portion can be reproduced faithfully, the deterioration of the image as a whole tends to be insignificant. Therefore, it is common to use narrow-band filters for the low-frequency regions and broadband filters for the high-frequency regions. Still images are amenable to two-dimensional and moving images to three-dimensional subband coding. Broadly speaking, the transform coding schemes can be considered instances of subband coding.

The main issues in subband coding are the subband analysis method, the design and characteristics of the filters, the bit allocation method, and the compression scheme within each band. The range of selections is extremely broad [3]. In particular, there are a number of candidates for the analysis and filter characteristics. The filters must not introduce distortion due to aliasing within inband analysis and synthesis.

Often the filter bank is implemented by quadrature mirror filters (QMF) used in layers [4]. The systematic design of QMFs and their regular structure makes them easy to use in practical applications. The symmetric short-kernel filter [5] has been shown to be more appropriate for video signal subband analysis than the QMF traditionally used in audio signal subband analysis, and is widely used. The advantage of the symmetric short-kernel filter is that the signal can be completely reproduced even if the filter is of low order. The wavelet transform, which has recently received much attention, can be seen as a type of subband filter bank method and interpreted as an expansion of short-time Fourier transform (STFT) [6]. The orthogonal wavelet bases are closely linked with the unitary two-band perfect-reconstruction QMF.

3.2.4 Entropy Coding

Entropy coding is an approach derived from information theory and categorized as a lossless coding method. It is a matter of how best to combine variable-length coding with the frequency of bit patterns appearing in the original data. There are many different combinations and coding methods.

For example, we can assign the following codes to the bit patterns appearing in the original data, beginning with the most frequently occurring bit patterns:

0, 1, 00, 01, 10, 11, 000, 001, 010, 011, 100, 101, 110, 111,

If we can determine the statistical properties of the original data, we can assign short-code bit patterns to the most frequent bit patterns in the source and assign long-code bit patterns to the least frequent bit patterns. This makes coding much more efficient.

This is an extremely lucid principle. In fact, the Morse code, which has a long history of use in telecommunications, is an embodiment of this idea, using a short signal (dot), long signal (dashes), and delimiters (two types). The frequency of the letters of the alphabet as they appear in English is as follows:

E, T, A, S, ...

Thus, the Morse Code was structured as follows:

E (·), T (–), A (·–), S (–·), ...

We can apply entropy coding to the ASCII code with a short signal (0) and a long signal (1) as follows:

E: 00001011 (**H) → 0
T: 00001011 (**H) → 1
A: 00001011 (**H) → 01
S: 00001010 (**H) → 10

However, it is not easy to solve the problem of determining the statistical properties of the original data, especially when these properties change over time (unsteady stochastic processes). In ordinary practical applications of image coding, the source data must be considered an unsteady stochastic process. In fact, the difficulties of creating accurate numerical models of source data are insurmountable barriers to practical implementation at this time.

At present, the efficiency of image coding is limited by the precision with which we are able to determine the statistical properties of the original data. In other words, usually we must be satisfied with a suboptimal level of optimality.

Two representative examples of entropy image coding are the Huffmann method and the arithmetic method. Both heavily process bit streams. In addition, the arithmetic coding method also makes frequent use of conditional decisions. These types of methods are not suitable for the DSPs that have been developed for the high-speed processing of vectors. This issue is discussed in more detail in Chapter 5, but for now let us emphasize that this issue must be considered when developing DSPs for image processing.

3.2.5 Vector Quantization

Quantization refers to the operation of making a group out of a set of continuous values and assigning a discrete value (code or bit pattern) to represent that group. Analog signals are continuous along the time and amplitude axes, so for digital signal processing both the time coordinate and the amplitude coordinate must be quantized and represented as digital signals. It is customary to refer to the quantization of the time coordinate as *sampling* and the quantization of the amplitude simply as *quantization*.

Vector quantization refers to the operation of making a group out of a collection of data (sample values) and assigning a discrete value (code or bit pattern) to represent that group. Whereas the term *quantization* is usually applied to quantization of analog values, in this case discrete values are being assigned to an aggregation of data. If each element of the aggregate data is expressed digitally, this maps several bit patterns onto a single bit pattern.

This is different from entropy coding in which bit patterns are one-to-one mapped onto bit patterns. It is readily apparent that vector quantization offers the great advantage of high coding efficiency. However, the problem for achieving these gains lies in finding practical algorithms for creating groups and mapping them with a minimum of processing. The algorithm proposed by Gray [7] surmounts this barrier, raising expectations that vector quantization will soon be practical.

Calculation of the distance between sample values is necessary for creating groups, and considerable calculation is also required for determining how to define the distance for the distance calculations. A rapid method of performing such calculations is therefore an important prerequisite for the application of vector quantization.

Note that since the distance calculations can be performed independently, this problem is eminently suitable for parallel processing. Furthermore, each set of image data is often suitable for vectorized processing, so it is possible to use a vectorized arithmetic processor for high-speed processing. Therefore, the problem of increased calculation size is not seen as a significant impediment to progress in vector quantization.

3.2.6 Motion Compensation

Motion pictures exhibit a strong correlation between the data in one frame and the data in the preceding and following frames. In practical terms, part of the image data may maintain its structure intact while its position moves over the background image data over the course of a number of frames.

If this motion can be expressed arithmetically and if the structure and its movement can be determined, then image data for several frames can be represented by a smaller volume of data. The quantity that expresses the particulars of the motion is called the *motion vector*, because vectors express size and direction [8].

To simplify the issue somewhat, an image may be divided into foreground objects and background. In motion compensation, the eye-catching movement of the objects against the motionless background is described as it occurs over an interval of several frames, thus reducing the amount of data. This is the most common method for the compression of motion data used in, for example, video games.

The most important issues of this method are (1) how to extract the motion vector, and (2) over what interval should it be extracted? As the interval is extended, the calculation grows dramatically, whereas it becomes impossible to correctly determine the motion vector if the interval is too short.

3.3 JPEG AND SHD IMAGES

This is an algorithm developed jointly by the International Electrotechnical Commission (IEC) and International Standards Organization (ISO) in 1991 [9] for the high-performance coding of still images, and is a basic combination of transform coding and entropy coding. The discrete cosine transform of 8 × 8 pixels is used for the transform coding. For entropy coding, Huffmann is used as the standard, and arithmetic coding is an option.

This algorithm specifies the compression ratio by only one type of parameter, called the *quantization step size*. When the step size is small, image resolution will be higher and the compression ratio smaller. The degree of compression that can be realized depends on the original image; the compression rate itself cannot be specified in advance. An unavoidable result of using DCT is that block distortion appears when high compression rates are used.

Figure 3.1 shows the structure of the algorithm. This algorithm can be implemented in approximately 2,000 lines of C code. Since still images need not always be processed in real time, this C program is very effective for many practical applications, although it can be time-consuming. Image coding and decoding can be performed on a PC or workstation without the addition of special-purpose hardware. However, considerable processing time may be required for large numbers of pixels. There are no constraints or specifications in the JPEG standard regarding the number of pixels.

The portion of overall processing in the JPEG algorithm that is devoted to calculations for entropy coding is far from small. Also, bit manipulation on the serial bit streams is not amenable to high-speed processing by the processor's arithmetic unit. Due to these factors, considerable processing time is required whether a PC, workstation, or DSP is used. Various vendors are therefore producing large-scale integration (LSI) circuits dedicated to JPEG coding. Use of such an LSI circuit on a PC or workstation speeds up the processing by at least an order of magnitude.

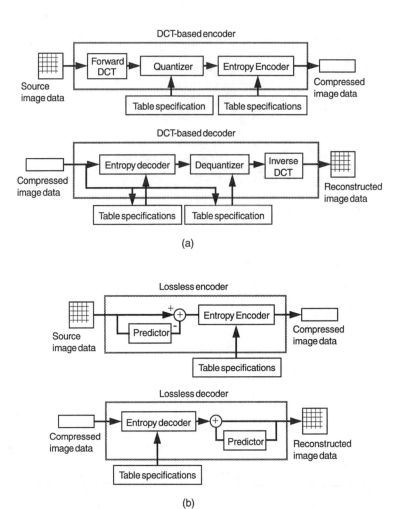

Figure 3.1 Block diagram of JPEG algorithm: (a) DCT-based encoder and decoder, and (b) lossless encoder and decoder.

Figure 3.2 shows an example of an SHD image compressed with JPEG. Figure 3.2(a) (see color insert) is an original image. The compressed image (compression ratio is 20:1) is shown in Figure 3.2(b) (see color insert). The signal-to-noise ratio (SNR) is about 31 dB. No degradation is visible at this compression ratio. The compression ratio of Figure 3.2(c) (see color insert) is 48:1 and the SNR is 26 dB. We can observe distortion in the sky of the image. Figure 3.3 shows the relationship between the SNR and the compression ratio for JPEG when applied to various images. Considerable variation between the images exists. We should be careful in using this algorithm in that it often results in pseudo-borders at compression ratios low enough to suppress observable block distortion. In particular, the compression of medical images with JPEG may cause fatal degradation even at the compression ratio of 2 to 1.

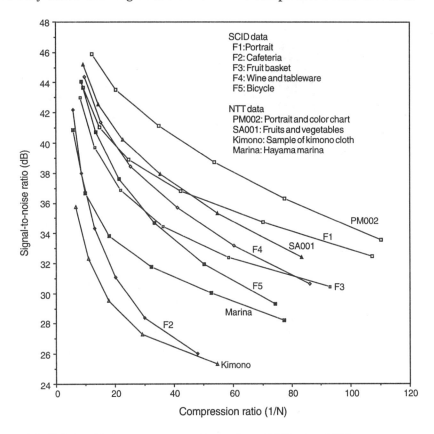

Figure 3.3 Relationship between compression ratio and SNR using JPEG.

3.4 MPEG AND SHD IMAGES

MPEG1 and MPEG2 are high-performance image-coding algorithms for motion pictures developed by the MPEG [10,11]. These algorithms include conventions for coding audio signals, but we will not discuss them here. For MPEG2, which was standardized most recently, there are 11 types of algorithms to accommodate different numbers of pixels and compression rates, as shown in Figure 3.4. Advanced Television (ATV, the American HDTV standard) is already included, and the European standard specifications will most likely be included. Therefore, this is likely to be a global standard in the near future.

Level ⟍ Profile	SP (Simple)	MP (Main)	SNP (SNR scalable)	SSP (Spatial scalable)	HP (High)
HL (High) [1920×1080×30 or 1920×1152×25]	Unused	MP@HL (ATV: US digital HDTV)	Unused	Unused	HP@HL
H1440 (High-1440) [1440×1080×30 or 1440×1152×25]	Unused	MP@H1440	Unused	SSP@H1440 (European digital HDTV)	HP@H1440
ML (Main) [720×480×29.97 or 720×576×25]	SP@ML (Cable TV)	MP@ML (Direct TV, Digital video)	SNP@MP	Unused	HP@ML
LL (Low) [352×288×29.97]	Unused	MP@LL	SNP@LL	Unused	Unused

Figure 3.4 Eleven specifications are defined in MPEG2. The numbers in brackets show the number of pixels in the horizontal direction multiplied by the number of pixels in the vertical direction multiplied by the number of frames per second.

This algorithm can be applied to an extremely wide range of image signals, including both interlaced and noninterlaced signals. At present, the algorithm has been specified only for the decoding side. This leaves considerable freedom in encoder design. MPEG1 was developed for coding motion pictures at the 1.5-Mbps bit rate available from CD-ROM. The image quality is lower than that of NTSC broadcast images. There is also a provisional version to raise the bit rate without altering the algorithm in order to raise the image to NTSC standards. This should not be confused with MPEG2, a completely revised set of standards, which was finalized at the end of 1993.

Broadly speaking, the MPEG algorithm consists of JPEG plus motion compensation. However, its development was compromised by an overly complicated decision-making process and a strict schedule, which led to a rather imperfect algorithm. Figure 3.5 shows the structure of the MPEG algorithm. This algorithm can be implemented in about 7,000 lines of C code. Since this coding algorithm is intended for use with motion pictures, it is oriented towards real-time processing. No matter how optimistically we consider progress in the processing power of PCs and workstations, there is no possibility that they will be able to process this algorithm in real time in the near future. A dedicated chip is required for real-time processing.

3.4.1 MPEG2 for SHD Moving Images—Is 150-Mbps Transmission Possible?

Figures 3.6a and 3.6b show the signal-to-noise properties and compression ratios when MPEG2 is used to compress SHD moving images. In these figures, the picture coding types I, B, and P represent the prediction method. The I frame uses only intraframe prediction, the P frame uses both intraframe and forward

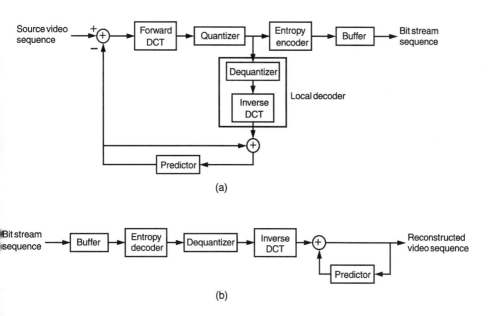

Figure 3.5 Block diagram of MPEG algorithm: (a) encoder, and (b) decoder.

interframe prediction. The *B* frame uses bidirectional interframe prediction. Figure 3.7 shows the relationship between the compression ratio and SNR of MPEG-encoded SHD moving images [12]. For reference, the performance of motion JPEG is also shown. Three sequences were examined. The sequence "fountain" contains multiple fountain nozzles, glittery water surface, and panning of the whole image, and thus resulted in the most severe drop in SNR. This result shows the very important information that we can get 40:1 compression ratio while achieving an SNR of about 40 dB. This will make it possible to compress SHD moving-image sequences, whose original data rate is 6 Gbps, to about 150 Mbps for economical transmission through one B-ISDN channel. Figure 3.8 shows every 18th frame from the sequences "swing," "marina," and "fountain" used for evaluating SNR properties. This figure is a gray-scale version of the original color images.

3.4.2 VLSI Development for MPEG2

An extremely high-speed chip with a high gate count is necessary for dedicated MPEG2 processing. Such chips are now being developed in all technologically advanced countries, and at ISSCC'94 there were reports that some had moved into the prototype stage [13]. These prototype chips are too large for existing mass production technologies, so it will be some years before a full-fledged mass-production version becomes available.

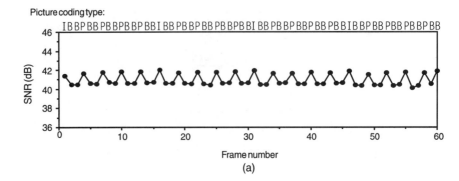

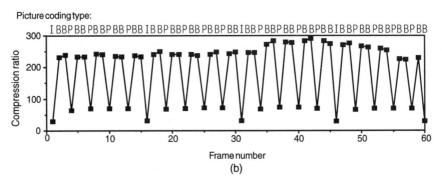

Figure 3.6 MPEG2 compression of SHD moving images (total number of frames is 60 for 1 sec): (a) SNR, and (b) compression ratio.

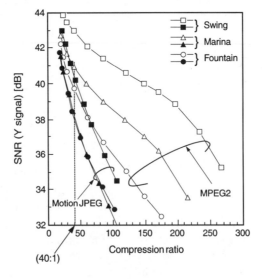

Figure 3.7 Compression ratio and SNR of MPEG2 for SHD moving-image sequences.

Figure 3.8 Image frames from the SHD sequences "swing," "marina," and "fountain" (up to down). Every 18th frame is shown (left to right: 0, 18, 32).

Since the MPEG2 decoder chip will be used for consumer HDTV equipment, the projected market is huge. Manufacturers are likely to engage in fierce competition to capture this market. Furthermore, the standards for MPEG2 are rather broad, so the specifications may be in excess of those required for practical applications. The increasing use of optical fiber means that the noninterlaced format will probably predominate for multimedia communications, so a subset or higher level version of the present standards is likely to receive special attention.

Under these conditions, problems are likely to surface regarding the hard-wired logic for fixed functionality of dedicated chips. Multi- and hypermedia devices require flexible functionality, and this is inimical to the idea of dedicated chips whose functionality is set in hard-wired logic. In other words, current developments in dedicated chips are merely aimed at solving the immediate problem of real-time processing to implement MPEG2; these chips are certainly not seen as a long-term solution. The authors predict that a system based on program logic, which can guarantee complete flexibility, will be needed for future multi- and hypermedia devices. This issue cannot be addressed by processors using conventional arithmetic elements.

3.5 SUMMARY

This chapter has addressed the fundamental algorithms for image compression and presented an overview of the JPEG and MPEG algorithms. We have focused on the result of applying these standard algorithms to SHD images. Our experience of applying JPEG to SHD images indicates that we must be very careful in employing it for certain applications because it often results in pseudoborders at low compression ratios. Coding algorithms for SHD motion images were also examined in terms of compression ratio and SNR. An important observation has been shown that we will be able to transmit MPEG moving images at the rate of 150 Mbps using the MPEG2 CODEC.

References

[1]   Nelson, M., *The Data Compression Book*, San Mateo, CA: M&T Publishing, 1992.

[2]   Rao, K.R., and P. Yip, *Discrete Cosine Transform*, San Diego: Academic Press, 1990.

[3]   Woods, J.W., ed., *Subband Image Coding*, Norwell, MA: Kluwer Academic Publishers, 1991.

[4]   Johnston, J., "A Filter Family Design for Use in Quadrature Mirror Filter Banks," *Proc ICASSP*, April 1980, pp. 290–294.

[5]   LeGall, D., and A. Tabatabai, "Sub-band Coding of Digital Images Using Symmetric Short Kernel Filters and Arithmetic Coding Techniques," *ICASSP88*, 1988, pp. 761–764.

[6]   Akansu, A.N., and R.A. Haddad, *Multiresolution Signal Decomposition*, San Diego: Academic Press, 1992.

[7]   Gray, A.M., "Vector Quantization," *IEEE ASSP Mag.*, Vol. 1, April 1984, pp. 4–29.

[8]   Musman, H.G., P. Pirsch, and H.J. Grallert, "Advances in Picture Coding," *Proc. IEEE*, Vol 73, April 1985.

[9]   Pennebaker, W.B., and J.L. Mitchell, *JPEG Still Image Data Compression Standard*, New York Van Nostrand Reinhold, 1993.

[10] LeGall, D., "MPEG: A Video Compression Standard for Multimedia Applications," *Communications of the ACM*, Vol. 34, No. 4, April 1991, pp. 46–58.

[11] CCITT Rec. H262, ISO/IEC 13818-2, "Generic Coding of Moving Pictures and Associated Audio," Committee Draft, Nov. 1993.

[12] Sawabe, T., J. Suzuki, and S. Ono, "Performance Analysis of MPEG2-Based Coding Algorithms for SHD Image Compression," presented at *Picture Coding Symposium*, Sept. 21–23 1994.

[13] Digest of Technical Papers, ISSCC'94, Session 4, February 1994.

Image Capture Systems   4

4.1 INTRODUCTION

High-quality digital image capture systems are indispensable for generating image media in the digital format. Image capture systems can be categorized into video cameras for moving images, and still cameras and scanners for still images. In general, scanners can capture images with higher resolution, but this takes a long time. On the other hand, CCD cameras have short capture times with lower resolutions. This chapter presents an overview of image capture technologies and practical systems, focusing on high-quality solid-state systems for SHD digital image capture.

4.2 VIDEO CAMERAS

4.2.1 NTSC/PAL/SECAM Video Cameras

Current manufacturing technologies allow 300,000 to 400,000 pixel high-performance CCD pick-up elements to be inexpensively manufactured on a one-third- to one-fourth-inch chip [1]. Video camera recorders (camcorders) using these devices are now common, particularly for the NTSC standard. Recently developed compact camcorders, which can record for as long as three hours using 8-mm or Hi 8 (high-quality 8-mm) video tape, offer quite good image quality, and are likely to become more popular. Image quality is obtained by extending only the frequency of the luminance signal, but this is relevant for the most common applications, such as home entertainment.

The prevalence of these high-performance inexpensive camcorders means that they can be used as input or editing devices for PCs configured for multimedia applications. Figure 4.1 shows an example of an inexpensive camcorder and editing system. Devices for making hard copies of still frames extracted from motion pictures obtained by this system have been marketed, but without

Figure 4.1 Photograph of a camcorder and a video editing system.

success, probably because the number of pixels is too small to provide an interesting picture.

Consumer-oriented video cameras almost without exception use just a single CCD, so color filters with a checkered color pattern are used over the CCDs. The quality of the resulting image signals depends on the color pattern of the filter and the color and luminance signal processing schemes. Recently, more emphasis on color reproducibility has prompted some companies to market consumer-model camcorders using three CCDs.

Most camcorders output composite VBS (video burst, synchronous) signals using PIN connectors. However, in the interest of improving image quality, S-terminals using DIN connectors have also been used; these allow luminance and color signals to be carried separately.

4.2.2 HDTV Video Cameras

In order to make HDTV video cameras more compact, CCDs have been developed that fit two million pick-up elements into one-half inch [2]. However, such CCDs are extremely expensive and have some weaknesses. The major factor influencing cost is the low manufacturing yield due to the large chip size.

This is a rather intractable problem, because making chips larger goes against the current trend in ULSI microengineering, which is to increase the number of elements per unit area. This has worked well for memory chips, such as CPUs and application-specific integrated circuits (ASIC), since gains in yield, processing speed, and integration (i.e., miniaturization) can all be realized at the same time.

However useful in other areas, this trend will not solve the problems of manufacturing CCD image pick-up elements. If the pick-up elements are made too small, the area for each pixel decreases with a corresponding decrease in light-gathering power. The amplitude of the resulting video signal will be low and the SNR will deteriorate.

Obviously, the fact that the amplitude of the video signal is low is not irrevocably linked to deterioration of the SNR, but in analog circuits such as those used in CCD pick-up elements, this is the inevitable result. This problem can be overcome by cleverly designing the electronic circuits, inserting a direct lens for every CCD pick-up element pixel, or providing some means of cooling the CCD pick-up elements to lower the thermal noise.

4.3 FILM CAMERAS

4.3.1 35-mm Film Cameras

Currently, the most common types of camera are 35-mm cameras, which accept a roll of 35-mm-wide perforated film. At the high end are single-lens reflex cameras which incorporate many recently developed automatic mechanisms, such as autofocus.

The theoretical maximum resolution of the highest resolution ASA 25 film is approximately $7,000 \times 7,000$ pixels. In fact, however, lens performance constrains practical applications to a much lower figure: $1,500 \times 1,700$ pixels. However, even this figure is not attained by many 35-mm cameras.

The upper practical limit of $1,500 \times 1,700$ pixels can only be attained by a professional skilled in both focusing and lighting techniques using a high-end camera. The average photograph taken by an amateur has a resolution of about $1,000 \times 1,000$ pixels.

The dimensions of the negatives are 35×24 mm. Images can be enlarged during printing up to 180×130 mm without image deterioration becoming apparent.

4.3.2 Brownie Film Cameras

Brownie roll film is 60 mm wide and has no perforations for feeding film, unlike 35-mm film. It is most frequently used for applications where the resolution of 35-mm film is inadequate, and is the widest roll film available for still cameras. The most common sizes are 45-, 60-, 70-, and 90-mm.

Professional photographers use Brownie film cameras more than any other. As mentioned earlier, the low resolution of 35-mm cameras is primarily due to inadequate lens performance, which can be overcome by increasing the aperture. While it is possible to use larger lenses with 35-mm cameras, the size of the film mount is more or less the same as a Brownie camera, so the larger lens is almost always combined with the higher resolution Brownie film in order to take full advantage of the more powerful lens.

For applications that require even larger film sizes, sheet film in 4 × 5- or 8 × 10-inch size is used. Cameras that can accommodate such film are rather large. In fact, the latter size is rarely used outside the studio.

4.3.3 70-mm Movie Cameras

The largest size film used in the motion picture industry is 70-mm, followed by 60-mm film. Note that 60-mm film is often scaled up and rerecorded onto 70-mm film for projection. IMAX, which uses 70-mm film, has the largest film surface (15 perforations).

Most motion pictures use 24 frames per second, but the Showscan format uses 60 frames per second. This faster frame rate is the temporal equivalent of higher spatial resolution, but its higher consumption of film is prohibitively expensive.

The 70-mm movie camera used by the authors is shown in Figure 4.2. This camera is intended for the optical measurement of objects in flight rather than cinematic uses; its maximum speed is 125 frames per second. It produces square images measuring 57 × 57 mm.

The maximum height of a frame captured by a cinematic camera does not exceed about 24 mm, so the resolution is no better than that of a 35-mm camera. A comparison of film types is given in Figure 4.3.

4.4 SCANNERS

4.4.1 Overview

Scanners are devices for converting images from film into electronic data, usually digital data. This is done by obtaining signals from sampling points on the

Figure 4.2 Photograph of 70-mm movie camera.

image fixed in the film. The sampling points are determined by an approximately square two-dimensional grid. While there is no special reason to limit sampling points to such a grid, there is no compelling reason to do otherwise.

Scanners may be categorized according to the method used for measuring signal values at the sampling points, the broadest difference being those that use single-point sensors and those that use multiple-point sensors. The latter may be further categorized into line sensors, which use a one-dimensional row of sensors, and area sensors, which use a two-dimensional array.

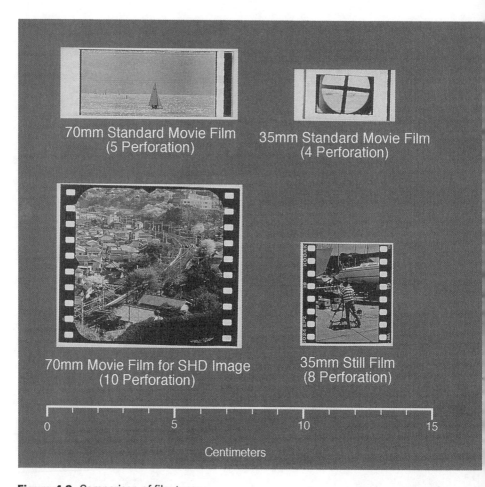

Figure 4.3 Comparison of film types.

If the number of sensors is less then the number of sampling points, the scanner must have a means of moving the sensors relative to the sample (i.e. scanning). Either the sensors or the film can be moved.

In most cases, mechanical means are employed for scanning. This means that the scanning speed is relatively slow, and scanning speed is the factor determining the processing rate. The scanning time is inversely proportional to the number of sensors. In other words, more sensors means shorter scanning time. The slowest scanner of all has only a single sensor.

Color scanners may be further categorized into those that scan R, G, and B signals simultaneously, and those that require one pass for each color. Obviously, the former type requires less time.

All signals obtained by the sensors are analog signals and must be converted to digital signals by A/D conversion. The modest bandwidth of these analog signals permits the use of converters having a precision of 12 to 16 bits. The bandwidth of these signals is not very high because of the long scanning time dictated by the mechanical means of scanning. However, the lower bits of the digital data always contain some sort of noise, and designers must always remain aware of this fact.

4.4.2 Drum Scanners

A drum scanner works by wrapping the original film around a transparent drum and slowly but accurately moving the position of the drum with respect to the sensor. The term *drum* refers to the transparent cylinder around which the film is wrapped. The scanning mechanism moves the sensor in a spiral along the cylinder. Figure 4.4 is a photograph of a typical drum scanner.

Figure 4.4 Drum scanner.

The advantages offered by drum scanners stem from the fact that only one sensor is used. This avoids the problems of differing characteristics for each sensor in a system with multiple sensors. Because only a single sensor is used, there are few constraints on its size, so an extremely sensitive but large sensor, such as a photo multiplier, can be used. Also, there are few problems with the light source, since a relatively low intensity spotlight can be used.

However, there are also several disadvantages that cannot be readily eliminated. The major problem is the time required for the data to be collected. This is a result of the scanning time, as discussed above, which can be shortened by revolving the drum faster. But even when the drum is revolved rapidly, the time remains quite long. Additional problems include low reproducibility and inaccuracy of the data collected, blemishes introduced on the film by the process of data collection (destructive read), and the like. All of these problems arise from the structure of the drum scanner.

More specifically, these problems stem from the use of a highly transparent liquid or powdered paraffin to fix the original film to the drum. Unfixed originals slip as the drum turns, lowering the resolution. Therefore, some sort of fixative must be used to obtain high resolution. The use of liquid or powdered paraffin requires considerable heat, and it is impossible to control the amount of liquid or powder used. Since the paraffin is not applied under uniform conditions, data reproducibility is suspect. Paraffin may appear to be transparent, but it does obstruct or reflect some light, which unavoidably shows up as noise between the sensor and the light source and compromises the accuracy of the data. Finally, the paraffin can only be removed by some physical means, which can damage the soft surface of the film. In fact, the effect of the paraffin layer on the film is so pronounced that even amateurs can tell the difference. The blemishes remaining after it is removed appear as countless small scars, which anyone can see.

These disadvantages limit the use of the drum scanner to specialists who are obsessed with high resolution. There is virtually no chance that this technology can be developed into a high-speed image input device with good color reproducibility usable by anybody.

4.4.3 Flatbed Scanners

In flatbed scanners, the original image is placed on a transparent plate, as in a copy machine, and the image data are then digitized by line sensors. The use of one- or two-dimensional line sensors means that either the sensors or the original image must be moved in a straight line. In some systems, the light source must be moved as well. The use of line sensors means that the distance that a flatbed scanner must travel is much shorter than is the case for a drum scanner. Consequently, the time required to capture image data is also much reduced.

Since the transparent plate holding the original image does not revolve, no fixative such as liquid paraffin need be used. Since the film is kept flat, the plate can be made of nonreflecting glass designed to prevent wasteful diffraction and the original can be held in place simply by placing the lid of the scanner over it. Placing the original on the scanner is therefore a much simpler task than with a flatbed scanner. However, the performance of such a simple mechanism is limited, and there are cases where liquid paraffin is used in order to obtain high-resolution data.

Whereas liquid paraffin is used with a drum scanner to prevent movement due to centrifugal force and vibrations caused by air currents, with flatbed scanners it is used to prevent the original from warping, and is thus not required for small originals.

Since a line sensor comprises a one-dimensional array of point sensors placed close together, many constraints are imposed on the shape of the individual point sensors. These constraints prevent improvement of the performance of the sensors. The properties of each sensor must also be identical. However, these problems are being solved by advances in LSI technology.

The real difference in the performance of flatbed scanners using line sensors and drum scanners using point sensors appears when film having a resolution and density index of 3.0 (i.e., a ratio of minimum to maximum density of 1,000) is used. This difference is due to the fact that the space between the individual sensors in a flatbed scanner is fixed, but the scanning pitch of a drum scanner can be varied mechanically. Furthermore, the photo multipliers used in drum scanners are much more sensitive than the CCD sensors of the typical flatbed scanner.

The advantages of the drum scanner are lost on most commonly used types of film up to the Brownie film size. Flatbed scanners using CCD line scanners give superior results for most film, including better SNR, and are far easier to use. Since data capture time is also faster for the flatbed scanner, current progress in automating the process of placing the original on the plate means that high throughput is possible.

The light source for flatbed scanners must be ingeniously designed to supply a uniform beam of light to all the line sensors. This can be done by making a single slit for each sensor and synchronizing the movement of the light source to that of the sensors.

4.4.4 Scanners for Motion Pictures

A device for obtaining video images (TV signals) from ordinary movie film is called a *telecine* device [3]. This device yields a video image for each frame of film. There are no technical problems with the video camera, since the resolution of the images captured on the film is more than adequate. The problem resides in the differing frame rates. Whereas most motion pictures have a rate of 24 frames per second, the video frame rate can be 30 (NTSC) or 25 (PAL, SECAM) frames per second. There are several ways of solving this problem.

No technical problems prevent film from being shot at 30 or 25 frames per second. In fact, this frame rate is often used for shooting television commercials because this produces quality images that are readily transferred to video.

The authors do not have access to an SHD video camera (since none has yet been developed), so we have developed a device for obtaining digitized motion picture data from 70-mm movie frames taken with a 70-mm movie camera [4]. This device is a combination of a high-resolution line scanner, a device for feeding the high-resolution movie film, and a halogen lamp as a light source. While our device resembles a Kodak Cineon, the Cineon was designed for 35-mm film and is not currently able to digitize data from 70-mm film.

The system developed by the authors is shown in Figure 4.5. This device is completely under the control of a workstation. The device itself is kept in a clean room to avoid dust and is monitored remotely. For a variety of reasons, this device is sensitive to minute vibrations, so it has been placed on an isolation bed that uses compressed air.

The frame-feed accuracy can be kept within 1 micron or less by use of both X and Y stages, which is enough to guarantee that not even one pixel per frame of displacement will occur. It takes a little over 10 minutes per frame to capture data for 2,000 × 2,000 pixel images. There are 12 bits each for R, G, and B. This data is sent from the workstation via fiber distributed data interface (FDDI) to a digital mass storage system (video tape–based optomagnetic disk) where it is stored.

The movie camera shown in Section 4.3.3 takes pictures at 60 frames per second. After the film has been developed, it is fed into our telecine device and digitized. This was shown on a CRT in 1992 as the world's first SHD video image. The frame memory systems that are required to display SHD video are discussed in Chapter 6.

4.5 ELECTRONIC STILL CAMERAS

If we combine a lens system, a CCD area sensor with necessary resolution, and an A/D converter with necessary accuracy, we can obtain a digital image capture system without scanning. The system is usually referred to as an *electronic still camera.*

As explained earlier, at the current state of technology, a maximum of two million picture-capturing elements can be constructed on a CCD. Such a CCD could be used in a still camera, but such products are currently nearly nonexistent. This is due to the low manufacturing yield and consequent high cost of CCDs having two million picture-capturing elements. In other words, the advantages of using an HDTV-grade CCD in a still camera are not great enough to justify the high cost. The only advantage of such a still camera would be the real time, or instant availability of images, which is already offered by Polaroid cameras. In addition to their high cost, such CCD still cameras are also large, heavy, and power hungry. Only rarely does the application justify their use.

An example of such a camera is the Scitex digital camera. Figure 4.6 shows the CCD sensor used in the camera. (The camera is shown in Figure 7.3

Figure 4.5 Photograph of film digitizer.

as a part of microscopy image digitizer.) The Scitex camera outputs 2,000 × 2,000 pixels with 8 bits on each of the R, G, and B channels. It takes 90 seconds to expose a picture; its sensitivity is equivalent to ASA 35 film. Therefore, it is most valuable when used as a high-speed scanner for capturing digital data from film. In practical terms, it is effective only for photographing still scenes.

The inconvenience of such electronic still cameras will be easier to understand if we contrast them with the Polaroid camera, which delivers a completed photo several minutes after the exposure is completed. Polaroid cameras closely resemble ordinary 35-mm single-reflex cameras. The only real differences are that they are slightly larger and require special film. But the special

Figure 4.6 Photograph of the CCD sensor used in Scitex's digital camera.

film is not noticeable once it has been loaded into the camera. In contrast, an electronic still camera requires a printer, magnetic storage device, batteries, monitor, and other peripheral devices, all factors that compare unfavorably with Polaroid cameras. Even if these components are combined into a single unit, as is done in some models, the device remains unwieldy.

Electronic still cameras are advantageous, however, when images must be rapidly transmitted to distant locations. (See Chapter 6 for the examples of microphotograph image transmission with electronic still cameras.) Current G3 and G4 standard facsimiles do not transmit images with halftones faithfully and cannot send color images. Color faxes are rather large. Not only does the electronic still camera do a better job of transmitting digitized images, but the received data can also be manipulated by the recipient.

If a CCD camera is to work as fast as an ordinary camera, there must be one image-capturing element for every pixel in the desired final image. This allows exposure times of less than 1 second, equal to ordinary cameras, to be achieved

However, if exposure times of several seconds or even several minutes are not a problem, it is acceptable to have fewer image-capturing elements than the final number of pixels, but the range of applications becomes more limited.

There are two types of CCD cameras that can capture images with a greater number of pixels than the camera has image-capturing elements: (1) a camera in which the image-capturing elements are infinitesimally displaced using, for instance, piezoelectric actuators, and (2) a camera in which there is infinitesimal movement of the light beam using prisms and other optical means. The corresponding mechanical movement obviously takes a certain amount of time, which contributes to the long image-capturing time for this scheme, as reported earlier.

A still camera requiring several seconds to several minutes to capture a picture can be used as a high-speed scanner, and several versions have been marketed. However, it is a fact that electronic still cameras have not yet been fully embraced by the marketplace due to the high cost of CCDs. The cost is due to the low manufacturing yields of defect-free CCDs containing this many elements, which in turn is due to the large surface area of the chips. Advances in ULSI manufacturing technology may make it possible to manufacture defect-free CCD image-collecting elements with high yield, so this problem is likely to be solved in the near future. The authors also predict that a low-cost practical system will be developed in the near future by hardware and software advances that allow the signal values of defective pixels to be inferred from adjacent pixels.

Once electronic still cameras surpass the resolution of conventional 35-mm cameras, their greatest attraction will be the ability to obtain digitized color image data directly, thus bypassing the use of color film. The long-standing problem of color film (i.e., the inevitability that film will deteriorate over time) will thus be completely eliminated. Images captured digitally can be stored with almost no deterioration whatsoever. Obviously, magnetic media also change over time, but much less than film. Furthermore, on magnetic media this problem is readily solved by transferring the digital data onto a new magnetic medium before they have reached the point at which they are in danger of being corrupted (i.e., at regular intervals, which can be quite far apart). The fact that digital images can be stored forever without deterioration will become a major advantage of electronic still cameras once the resolution becomes better than that of 35-mm film cameras.

4.6 SUMMARY

Current image capture technologies and future trends have been discussed focusing on practical capture systems. Conventional video and film cameras covering sizes from 35- to 70-mm have been described. Technologies related to scanners have been discussed from the viewpoint of obtaining digital high-quality image data. It is possible to capture high-resolution images using conventional scanner technologies if the required long time to obtain the captured data is acceptable. However, advances in VLSI technologies will allow conventional low-speed scanners to be replaced by electronic still cameras with large defect-free CCDs. This technological advance will make possible real-time digital image capture systems, even SHD video cameras.

References

[1]   Suzuki, J., et al., "A 1/4-inch 250k Pixel IT-CCD Image Sensor," *IEEE Trans.*, Vol. 39, No. 3, Aug. 1993, pp. 392–397.

[2]   Yonemoto, K., et al., "A 2-Million Pixel FIT-CCD Image Sensor for HDTV Camera System," *ISSCC Dig. Tech. Papers*, 1990, pp. 214–215.

[3]   Sandbanl, C.P., Chap. 10 of *Digital Television*, Chichester, England: John Wiley & Sons, 1990.

[4]   Kashiwabuchi, K., I. Furukawa, and S. Ono, "Film Based Motion Picture Digitizing System for Super High Definition Images," *Proc IS&T/SPIE Symp. on Electronic Imaging Science & Technology '94*, Vol. 2173–18, Feb. 1994.

Signal-Processing Technology ■5

5.1 INTRODUCTION

When digital SHD data are to be transmitted, stored, edited, image-enhanced, or embellished with special effects, systems capable of processing huge data sets at high speed are indispensable. In this chapter, we will consider the technologies necessary to implement such signal-processing systems. First, we will turn our attention to the performance and functional requirements. Next, architectures for these systems are discussed, focusing on those based on DSPs, whose flexibility makes possible the implementation of many types of algorithms. As a promising architecture for SHD image processing, a parallel, loosely coupled DSP system is introduced. The NOVI-II parallel signal-processing system is explained as an example of such systems.

5.2 REQUIREMENTS FOR SHD IMAGE PROCESSING

Figure 5.1 illustrates the signal-processing flow for all-digital SHD image and video signals. Digital signal-processing functions can be divided into preprocessing steps such as noise reduction and image enhancement, compression, and postprocessing for reproducing high-quality images. The basic algorithms for these signal-processing functions are not essentially different from those that have been used with traditional image-processing systems. However, because the volume of data is an order of magnitude greater, correspondingly higher performance is called for. In fact, roughly eight times more performance is required to process SHD images as compared to current broadcast TV images. This is because SHD images have four times as many pixels and twice the frame rate. Moreover, depending on the application, extremely high precision may be required, so even higher performance may be demanded in the future. The complexity of processing required depends on the service needs of the system as a whole.

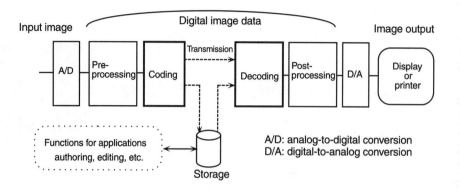

Figure 5.1 Signal-processing flow for SHD images.

Signal processing for compression is generally used to deliver high-quality services at the lowest possible cost. For instance, by reducing the volume of data without degrading the image quality, data can be transmitted in a shorter time or it can be stored in a smaller memory. Unfortunately, the real world is very complex. Let us take a look at transmission time. Image distribution systems must minimize delays throughout the entire system in order to reduce the access time (the time between a request issued by a terminal and the display of the image at the terminal). The system's overall delay is a composite of the database access time, time to decode the image data, and transmission delay time. In general, complex encoding and decoding increases the compression factor and reduces the transmission delay, but at the expense of increased processing time for encoding and decoding. Furthermore, the complexity of the decoder is increased. Dependence on overly complex signal processing may not be the optimal system design decision. If B-ISDN fulfills its promise of delivering high-speed circuits (150 Mbps or better) at low cost, it may be preferable to perform simple processing and take advantage of the high-speed channel, rather than requiring high signal-processing power to compress or decompress SHD image data. The designer will have to take into consideration many aspects of the overall system when making decisions about how to apply signal processing.

Below, we will discuss requirements for processing performance and flexibility.

5.2.1 Estimated Processing Power

Let us roughly determine the arithmetic performance required to process high-quality images with a resolution of 2,000 × 2,000 or better. The raw data making up a single basic SHD image are 2,000 × 2000 pixels × 8 bits × 3 colors, for a grand total exceeding 12 MB. If the effective transmission rate is 100 Mbps, this image can be transmitted in about a second. If we let T represent the time required to display an image from the moment it is requested by the terminal (the access time), then $T < 1$ second is the minimum requirement for a real-time service. T includes, in addition to signal-processing time, database access time and transmission time, so we had better make certain that we can complete the processing in less than 0.1 second. Now, if we want to use moving images with a frame rate of 60 frames per second, the processing of each frame must be completed in less than 1/60 of a second.

Many popular image-encoding algorithms are based on combinations of fundamental techniques such as DCT and vector quantization (VQ). Motion compensation (MC) is also used for moving images. Figure 5.2 shows the relationship between image resolution and the signal-processing power required to perform the compression of moving images in real time. The power required is represented by billions of (floating point) operations per second (Gops or Gflops). The resolution is taken as the number of vertical lines, with the simplifying assumption that the image format is square. The algorithms are considered to be applied to the luminance component of the images, along with their subsampled chrominance component. Figure 5.2(a) shows the performance required to apply DCT with an 8×8 block size, while Figure 5.2(b) addresses MC, using a full search, a block size of 4×4, and a search region of plus or minus 16×16 pixels. As these graphs show, the application of DCT in real time to SHD moving images of 60 frames per second calls for 15 Gflops, and MC calls for an additional 1 Tflops of processing power. Note, though, that this estimate is based on the assumption that calculations must be performed for all pixels, and it is thus best considered an upper limit. Actual encoding algorithms are adaptive, with SNR varying according to the characteristics of the image, and the computational expense of achieving a given SNR is also variable, depending on the image data. We have applied JPEG, a standardized encoding algorithm, to a number of still SHD images in order to determine the actual computational expense; the results are presented in Figure 5.3. Picture numbers shown in Figure 5.3 correspond to the standard SHD images in [1]. As the figure shows, approximately 50 Gflops are required to perform JPEG compression on moving images at the rate of 60 frames per second. When MC techniques were employed in JPEG compression, at least 150 Gflops were required.

Sawabe investigated signal-processing power requirements of MPEG2-based coding algorithms for SHD moving-image compression [2]. MPEG2 is a moving-image encoding standard recently developed for video discs, digital video telephone, and video conferencing [3]. The amount of computation required for MPEG2-based compression depends on the algorithm and parameters for MC, contents of images, compression ratio, required SNR, and other parameters.

Table 5.1 shows the processing performance required for MPEG2 coding and decoding when applied to a 2048×2048 SHD moving-image sequence. The algorithm used is an extended version of the so-called MP@ML; the level was

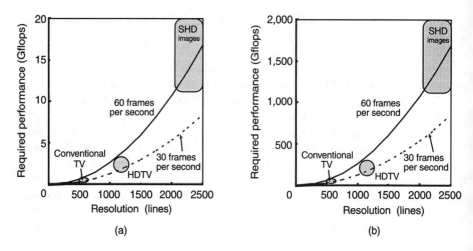

Figure 5.2 Computing performance required by (a) two-dimensional DCT, and (b) MC.

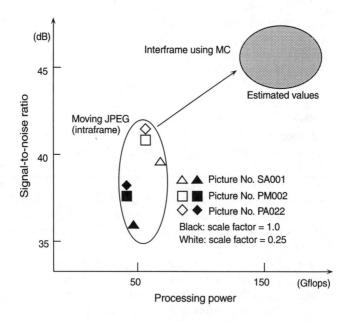

Figure 5.3 Measured processing power and SNR for SHD image compression.

extended to cover 2048 × 2048 × 60. The number of digital signal-processing operations in coding and decoding was measured for an actual image sequence digitized from 70 × 70-mm movie films. In the calculation, full-search and tree-search MC schemes were compared. Both schemes achieved 40:1 compression at an SNR value of about 40 dB. The performance required for real-time MPEG2 coding and decoding for moving SHD images was estimated to be about 300 Gflops. For reference, it is estimated that the number of arithmetic operations for the standard MP@ML (720 × 576 pixels, 30 frames per second) is about 1/8 of the extended MP&ML for SHD moving images.

These estimates clearly demonstrate that even considering foreseeable advances in the latest VLSI technology, large-scale parallel processing systems will be required to implement DSP systems for processing SHD images.

Table 5.1
Number of Operations and SNR for an SHD Moving Image

MC Scheme	# of Operations in Encoder per Second	# of Operations in Decoder per Second	SNR at 150 Mbps
Full search	661 billion	22 billion	42.84 dB
2 stage tree search	261 billion	22 billion	42.83 dB

5.2.2 Programmable DSP or ASIC?

In order to determine the best algorithm for image compression, many trial-and-error tests will have to be performed. This, in turn, indicates that programmability is required to guarantee flexibility in terms of the signal-processing algorithms. In particular, during the initial research and development phases, programmable DSP-based systems will be indispensable. In general, processors deliver lower performance than hardware. Thus, some architectural sophistication, such as parallel processing, will be needed to deliver real-time performance from a programmable system. As we discussed above, even foreseeable advances in LSI technology will not deliver adequate performance from a single DSP, so multiple DSPs will have to be configured into a parallel system.

On the other hand, in order to guarantee a large market for services, the signal-processing component must be made available at low cost. Application-specific LSI (ASIC) technology can help accomplish this goal by reducing chip counts. A real-life example of how ASICs can do this is the development of dedicated LSIs for image processing. Development of these dedicated LSIs, which pointed the way to a viable market, began at about the same time as CCITT (Consultative Committee in International Telegraphy and Telephony) efforts to standardize H.261 first got under way. First, chips that could execute DCT, the fundamental algorithm for encoding, appeared on the market. Next, LSIs were developed that offered other important functional blocks for encoding: MC, variable-length encoding, and quantization. Finally, a chip set appeared, which implemented the H.261 CODEC [4]. However, as the complexity of the encoding method continues to increase, the need for flexibility vis-à-vis individual applications also continues to increase, and it becomes difficult to implement all processing operations flexibly enough simply by combining an ASIC chip set. Therefore, programmable DSPs will continue to play a significant role, even when ASICs are used.

5.3 DIGITAL SIGNAL PROCESSOR

5.3.1 DSP Evolution

DSP evolution can be viewed as consisting of three stages. In the early days, DSP architectures were designed to support the fast calculation of the strings of multiplications and additions, or vector calculations, needed to implement digital filters and the like for signal processing. The mission of these early DSPs was to perform signal processing in the audio band, delivering performance that was beyond the reach of general-purpose microprocessors. The principal applications were digital filters, modems, and voice CODECs [5]. This phase extended from 1980, when the first DSP was developed, to about 1986. LSI technology was at that time at the 3-micron level.

The second stage of DSP evolution started in 1987, just when ASICs first appeared. These more sophisticated second-generation DSPs offered longer words, richer addressing functionality, and larger memories, and began to find their way into real-time control, audio processing and synthesis, high-functionality modems, and certain image-processing applications. LSI technology in those days was mainly working with 1.5-micron rules. JPEG encoding can be performed with these second-generation DSPs.

Third-generation DSPs have 32-bit-wide arithmetic paths and thus operate on the same 32-bit floating-point format as do general-purpose microprocessors. The announcements, beginning around 1988, of these third-generation devices fabricated with 0.8- to 1-micron rules heralded wider applications and higher performance. At the same time, though, general-purpose numeric processors of the time were endowed with the same dedicated arithmetic units as DSPs and with addressing functionality suited to signal processing, and were beginning to overtake the DSP in terms of performance. Those DSPs, called video signal processors (VSP), could be combined in tightly coupled multiprocessor systems to perform video signal processing, and are also considered third-generation DSPs.

In recent years, DSPs with multiple communication ports suitable for implementing loosely coupled multiprocessor systems have been announced, and this phenomenon may be the beginning of the fourth generation of DSPs [6]. Figure 5.4 illustrates DSP evolution in terms of performance and application range. DSPs are always in competition with both custom LSIs and general-pur-

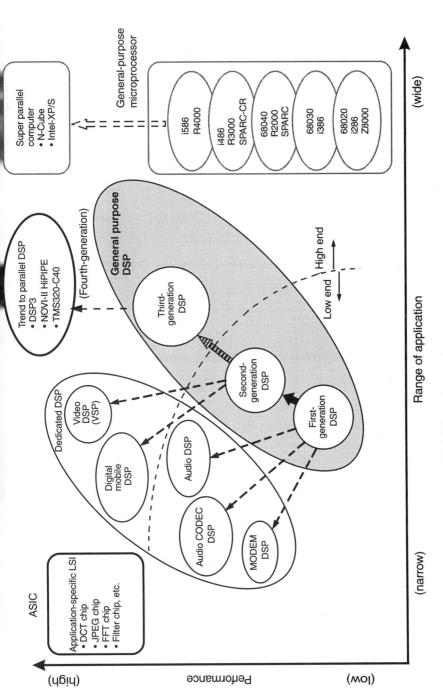

Figure 5.4   Evolution of the DSP: comparison with ASICs and general-purpose processors.

pose LSIs. Their success to date has resulted from their combination of high speed and programmability.

5.3.2 Basic Architecture of the DSP

The DSP architecture is a design that gives top priority to the continuous streams of multiplications and additions that are the hallmark of digital signal processing. The Harvard architecture is combined in a pipeline manner to supply data continuously to the arithmetic unit. The Harvard architecture separates the data and program memory spaces, along with their buses. This leads to reduced efficiency of memory usage, but allows required data to be efficiently delivered to the arithmetic unit; the organization places top priority on the arithmetic unit. This architecture is depicted in Figure 5.5.

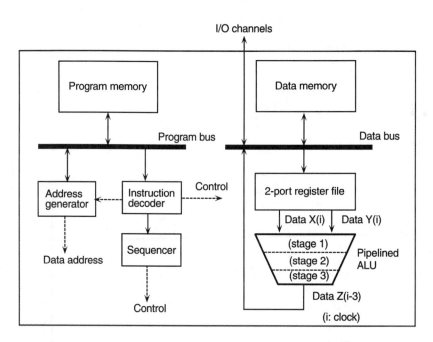

Figure 5.5 General DSP architecture.

The arithmetic unit is divided into small stages to allow the pipeline to operate. The throughput of the arithmetic unit's pipeline can be increased by reducing the pitch of the pipeline, but because this complicates arithmetic unit control, distribution of data from the register file, and address generation, single-chip DSPs are practically limited to pipeline pitches of several nanoseconds. Because digital signal processing makes heavy use of sums of products, all DSPs are designed with arithmetic units that shine at this operation.

5.3.3 Video Signal Processors

Here we will turn our attention to VSPs, which we dubbed the third-generation DSP in Section 5.3.1. Table 5.2 demonstrates some examples of VSPs announced during the past few years. Figure 5.6 shows the DSP architecture as a representative example of such a device [7]. The figure illustrates how the memory, address generator, bus organization, and data paths surround the arithmetic unit. The organization of these components typifies the third-generation DSP architecture. Other architectural features of note are:

- Strong two-dimensional addressing mode to support image processing;
- Built-in pipeline processing for distance calculations (i.e., calculating the difference of a pair of vectors);
- Large on-chip memory;
- Suited for use as processing element in a tightly coupled parallel processing system.

Table 5.2
Commercial VSPs

Manufacturer	Mitsubishi	NEC	NTT	IIT
Product name	DISP	S-VSP	IDSP	Vision Processor (VP)
Technology	1-mm CMOS[*]	0.8-mm BiCMOS[†]	0.8-mm BiCMOS	1-mm CMOS
Package	177-pin PGA	132-pin PGA	280-pin PGA	144-pin PGA[‡]
Transistor count	538 kTr[§]	1	130 kTr	910 kTr
Power dissipation	7W (typical)	1.8W (typical)	1W (typical)	1.6W (typical)
Instruction execution time	50 ns	4 ns	40 ns	20 ns
Data word length	24-bit fixed point	16-bit fixed point	16-bit fixed point	8/16-bit fixed point
Multiplier/ALU	$24 \times 24 \to 47$ bits	$16 \times 16 \to 35$ bits	$(16 \times 16 \to$ 24 bits) $\times 4$	$(16 \times 16 \to$ 32 bits) $\times 4$
	24-bit ALU	16-bit ALU	(16-bit ALU) $\times 4$	(8-bit ALU)[‖] $\times 16$
On-chip memory for instruction	512 wd $\times$ 48 bits (RAM)	32 bits $\times$ 1 kwd (RAM)	32 bits $\times$ 512 wd (RAM)	Instruction cache

Table 5.2 (continued)

Manufacturer	Mitsubishi	NEC	NTT	IIT
On-chip memory for data	512 wd × 24 bits 2 dual-ported RAM	5.5 kwd × 16 bits (total) dual-ported RAM	512 wd × 16 bits 5 dual-ported RAM	2.9 kwd × 8 bits (total)
Addressing	Two-dimensional			
External interface	Instruction port	Data port × 3	Data port × 3	Data port
	Data port	Host microprocessor	Serial port	Command port
				Optional microcode port
DMA	Two-dimensional DMA#	–	3 modes	1 mode
Released	'89	'91	'91	'91

Notes:
* CMOS is an acronym for complementary metal oxide semiconductor.
† BiCMOS is an acronym for bipolar complementary metal oxide semiconductor.
‡ PGA is an acronym for pin grid array.
§ kTr is an acronym for kilotransistors.
|| ALU is an acronym for arithmetic and logic unit.
\# DMA is an acronym for direct memory access.

The design's emphasis on application of the part in parallel processing is evident, such as in the organization of the data bus that tightly couples the memory and the arithmetic unit. For example, with the IDSP, four banks of memory may be connected to four arithmetic units, which allows 8 words of data to be simultaneously presented [8]. Figure 5.7 shows an example provided by IDSP of memory and arithmetic unit organization. We have added to the figure a technique for efficiently executing MC using this organization. The S-VSP uses an exceedingly small pipeline pitch, so that performance can be increased by connecting dedicated memory to very-high-speed arithmetic units. These VSPs were basically developed in order to implement MPEG encoding.

These chips are not suitable, however, for tightly coupled parallel systems with more than several elements. This is because the performance will saturate due to the complexity of shared memory control and due to saturation in the data transfer rate between elements as the number of elements increases. Thus, processing SHD images calls for a loosely coupled system that can incorporate at least 100 DSPs, based on such multiprocessor DSPs as the T9000 Transputer, or the TMS-C40. The specifications for these parts are shown in Table 5.3. In

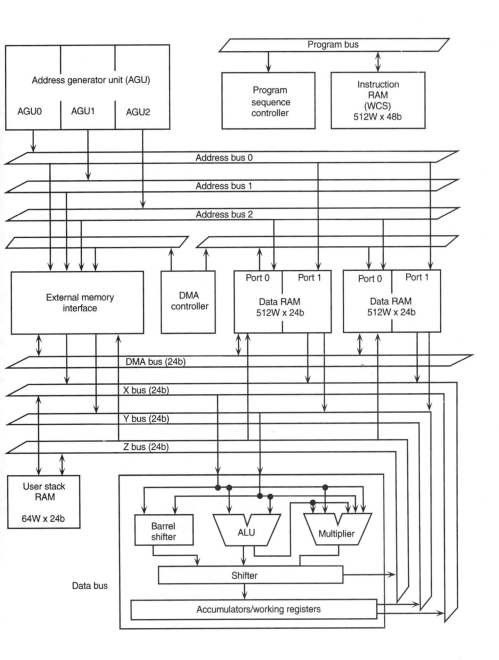

Figure 5.6 Example VSP design (from IDSP) [7].

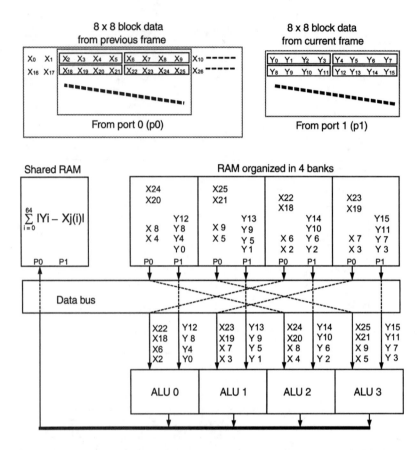

Figure 5.7 Data distribution used for distance calculations in MC (example from IDSP).

the next section, we will present the SHD image-processing system developed in our laboratory, which has such a multiprocessor organization exactly.

Table 5.3
Example of Recent Loosely Coupled DSP and Multiprocessor Products

Item	TMS320c40	T9000 (H1-50)	(DSP96002)
Speed	25 MHz	50 MHz (internal)	33 MHz
Communication links	6×20	Mbps (bidirectional link)	4×20 Mbps (bidirectional link)

Item	TMS320c40	T9000 (H1-50)	(DSP96002)
Peak performance	25 Mips	150 Mips	16.5 Mips
	50 Mflops	20 Mflops	49.5 Mflops
Memory (on-chip RAM)	1 kwd × 2, program cache 128W, (ROM 4kw)	16 KB cache (program and data)	program 1 kwd data 0.5 kw × 2
Programming language	Assembler C	Occam 2 Fortran, C, Ada	Assembler C
Program development and debug environment	In-circuit emulator (ICE) adapted for multiprocessor, real time operating system (OS)	Evaluation board, real time OS	Internal ICE functionality, real time OS
Word length	32-bit floating point (two's complement)	IEEE 754 (32 bits)	IEEE 754 (32 bits)
LSI technology	0.8 micron CMOS	0.8 micron CMOS	1 micron CMOS

5.4 THE NOVI-II PARALLEL SIGNAL-PROCESSING SYSTEM

The NOVI project commenced in the NTT Transport Processing Laboratory in 1987, with the goal of achieving parallel signal-processing performance in the range of trillions of floating operations [9]. Commercial processors of this type have very recently appeared, and a parallel system built around them has been announced [10]. In this section, we will present the NOVI-II as an example of an advanced parallel signal-processing system.

The NOVI-II is a multicomputer architecture DSP system. The current version consists of 128 processing elements (PE) and delivers a maximum of 15 Gflops. The overall organization is shown in Figure 5.8. Figure 5.9 shows the organization of a single PE. The T800 Transputer [11] provides the interprocessor communications, special vector processor chips provide excellent arithmetic performance (120 Mflop), and the probe chip built into each PE provides hardware assistance for debugging the parallel programming environment. The system is programmed in high-level languages such as Occam [12] or C [13]. The hardware probe simplifies programming and debugging by giving the programmer a clear view of multiple processes even as they execute in parallel.

Figure 5.10 shows a photograph of the NOVI-II. While the NOVI is certainly applicable to general signal-processing tasks, at the moment it is harnessed to our experimental SHD image system.

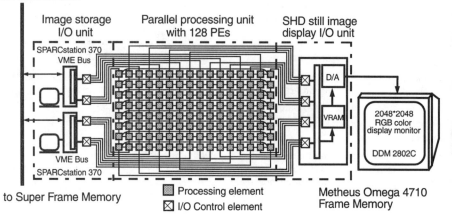

Figure 5.8 Block diagram of NOVI-II for SHD image processing.

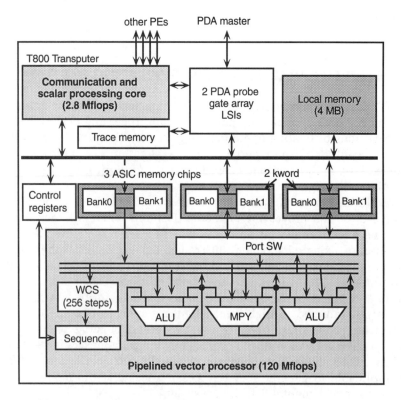

Figure 5.9 NOVI-II processing element configuration.

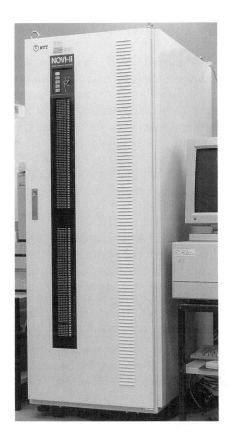

Figure 5.10 Photograph of NOVI-II system.

5.4.1 Parallel Processing Architecture of the NOVI-II

The most promising architecture for SHD video processing is a multicomputer in which the elements do not share memory. There are two principles behind this thinking.

1. *Maximize PE performance.* Although it seems rather obvious that individual PEs must be high-performance in order to maximize the performance of the system as a whole, it is an extremely important principle. In particular, when operations like signal processing, which has a highly regular structure, are performed on blocks, a massively parallel system of thousands to tens of thousands of simple, lower performance processors may be swamped by communications overhead, mitigating the possible performance gain. To restate, in order to deliver

high performance, a parallel signal-processing system must have large granularity.

2. *Minimize shared resources.* Even if a high-performance PE is used, the interprocessor communications must be minimized in order to maximize the overall system performance. In general, sharing fewer resources minimizes the interprocessor communication. Of course, the down side is that total resources are increased. In the case of PE memory, however, we can expect advances in VLSI technology to come to our rescue by continuing to bring down the cost of memory. On the other hand, if the number of PEs increases, the diverse communications requirements make it difficult to guarantee adequate inter-PE communications bandwidth.

For these reasons, the NOVI-II architecture features a pipeline based on multiple arithmetic units, large local memories, strong interprocessor communications support, and high-performance PEs. In other words, we have implemented our parallel DSP system as a loosely coupled message-passing multicomputer. Of course, the size of the individual PE memories as well as the inter-PE communications bandwidth is large enough to handle the huge volumes of data inherent in SHD images. NOVI is as yet a prototype and, as we will discuss in the next section, does not yet possess the performance required to process SHD video in real time. However, we believe that the performance increases required for SHD high-end video processing systems can be achieved with this sort of architecture.

5.4.2 NOVI-II Performance

In general, the sustained performance of a parallel system is considerably less than its peak performance. We have developed a library of parallel software tools for SHD video processing, particularly compression, and measured their actual performance. These tools are written in a version of C that takes advantage of the parallel constructs offered by the Transputer [13]. A dedicated cross-assembler is used to program the vector processor (VP); the microcode emitted by the cross-assembler is loaded into the VP from the Transputer. For comparison, we have also coded the same tools in Fortran and run them on a single CPU, Cray-2. Note, though, that the Cray-2 uses a 64-bit floating-point format.

Table 5.4 shows the actual performance measured for a number of VSP calculations in these environments. The input data were $2,048 \times 2,048$ full-color digital SHD images (8 bits per R, G, B). The parallel performance shown in the table was estimated by the following simple procedure.

1. Image subdivision, whereby image data are divided into 128 blocks, and each PE operates on a single block. Each PE has 128 MB of local memory.

2. Each operation is carried out within a PE. When data from another block are required, the Transputer's communications links are used to transfer it.
3. After all operations on blocks are completed, the 2048 × 2048 SHD image is reconstituted from the blocks. Details of the operations are given in Table 5.5.

Table 5.4
Measured Processing Time for Basic Algorithms for SHD Image Compression

	NOVI-II With 128 VPs	NOVI-II Without Any VP	Cray-2 (single CPU)
Conversion of RGB to YIQ*	38 ms	767 ms	15,509 ms
Conversion of YIQ to RGB	52 ms	1,096 ms	19,675 ms
2D DCT	21 ms	2,122 ms	15,274 ms
Subband 1 (32-order QMF)	95 ms	3,451 ms	
Subband 1 (32-order QMF)	97 ms	3,788 ms	29,014 ms
Subband 2 (reconstruction filter)	51 ms	676 ms	
Subband 2 (reconstruction filter)	32 ms	680 ms	5,794 ms
VQ	364 ms	34,571 ms	82,033 ms
MC	2,473 ms	183,817 ms	1,870,703 ms

*Note: YIQ is a color signal representation of NTSC.

Table 5.5
Compression Operations Used To Measure NOVI-II Performance

Operation	Details
RGB-YIQ conversion	This operation converts an RGB signal into a YIQ signal. The YIQ representation is more suitable for application of compression algorithms.

Table 5.5 (continued)

Operation	Details
Two-dimensional DCT	This is a two-dimensional DCT with a block size of 8 × 8 and suitable for a continuous pipeline. It takes a little time to transfer the DCT blocks.
Subband 1	This is a 32-order QMF filter. It takes time to transfer the block environment data required to perform two-dimensional filtering. We compared two methods.
Subband 2	This uses a fifth-order low-pass and a third-order high-pass filter [14].
VQ (vector quantization)	Block size = 8 × 8, codebook size = 256 × 64 words, full pass filter.
MC (motion compensation)	Block size = 4 × 4, search region = plus/minus 16 pixels, full search algorithm. The Vp can do distance calculations very quickly, but the large search region forces time to be spent transferring data.

Table 5.4 shows that the presence of the VP improves performance significantly, but with significant variation depending on the operation considered. This is because some operations are less able to keep the VP pipeline full and because the inter-PE communications overhead differs for each operation. For instance, the VQ encoding operation, which uses an 8 × 8 block size, a 256 × 64-word code book generated using the full-search method, is able to take full advantage of the VP pipeline. For this reason, VQ code book generation reaches nearly 10-Gflop performance. On the other hand, subband filter processing must exchange data from other blocks in a two-dimensional fashion, giving rise to significant communications overhead. The table illustrates the performance for subband Methods 1 and 2, which differ in their implementation of data exchange. Method 1 first sends only the data needed to process rows to each PE and when row processing is completed resends the data needed to process columns. In Method 2, all data needed to process both rows and columns are first sent to the PEs. Although Method 1 gets by with fewer floating-point operations, Method 2 demands less communication. Thus, when a VP is present Method 2 finishes faster than Method 1.

Figure 5.11 shows, for each operation, the portion of total execution time during which the VP is operating. The "total processing time" and "VP executing time" are derived by dividing the time for the operation by the number o

floating-point operations. The major processing overhead is data exchange between memory and the VP.

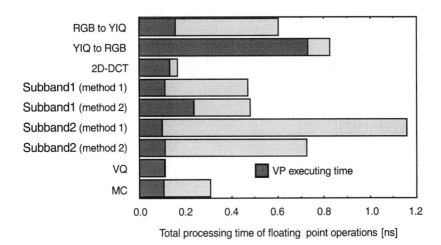

Figure 5.11 VP execution time.

5.4.2.1 JPEG Compression of SHD Images

We implemented the JPEG standard for still image compression, and applied it
in the NOVI-II system to SHD images [15]. The processing times are shown in
Table 5.6. JPEG is a combination of compression, largely based on DCT, and
Huffman encoding. It is shown that using the VPs greatly improved the processing time for compression (about 50 times faster). On the other hand, they improved Huffman encoding by about a factor of 5. Although the computational
expense of the Huffman encoding portion of the algorithm varies depending on
the compression ratio, we can encode a single color SHD image in about 150 ms
with the JPEG algorithm on the present NOVI-II.

Table 5.6
Processing Time for JPEG Encoding

	NOVI-II HiPIPE With 128 VPs	NOVI-II HiPIPE Without Any VP
Conversion of RGB to YUV* + 2D-DCT + zigzag scan + quantization	94 ms	5,001 ms

Table 5.6 (continued)

	NOVI-II HiPIPE With 128 VPs	NOVI-II HiPIPE Without Any VP
Huffman encoding	50 ms	332 ms
Scale factor = 1.0 (compression ratio = 53.7)		
Huffman encoding	78 ms	438 ms
Scale factor = 0.25 (compression ratio = 20.4)		
Total processing time (scale factor = 1.0)	144 ms	5,333 ms
Total processing time (scale factor = 0.25)	172 ms	5,439 ms

*Note: YUV is a color signal representation recommended by the International Telecommunication Union (ITU).

In order to process 60-frame-per-second SHD video, the performance must be increased by about one order of magnitude. However, the VPs are of little help in the type of bit operations called for by Huffman encoding, so architectural innovations will be necessary to execute the Huffman part of the algorithm at high speed, as will be discussed in the next section.

5.4.3 Towards Real-Time Processing of SHD Video

As our discussion to this point has shown, while NOVI-II can compress SHD still images at high speed, it is not yet up to the task of processing moving images in real time; further performance improvements are needed. Let us take a look at the architectural aspects that offer room for improvement.

Advances in VLSI technology will soon make it possible to implement the VP with 0.3-micron rules, yielding a tenfold increase in density and a fourfold increase in clock frequency. This will allow memory, formerly placed off-chip, to be moved on-chip, for a total increase in performance of about 10 times. At this point, inter-PE communications will be the performance-limiting bottleneck. This effect can be countered by adding a very large memory to each PE in order to reduce inter-PE communications, and by increasing the bandwidth of the inter-PE communications links. Let us consider how much bandwidth is needed.

First, data must be input at 720 Mbps in order to process SHD video in real time. The parallel DSP system must be able to maintain inter-PE data transfers while inputting and outputting this 720-Mbps data stream. The three-dimensional cube network is considered a promising candidate for a topology suited to supporting this type of communication [16]. An example of this network topology is shown in Figure 5.12. This topology supports both data parallel and pipelined PE models, and has large input and output bandwidth. In order to process SHD video with 128 PEs organized as a three-dimensional cube, each PE would need six links with at least 50-Mbps bandwidth per link. The design of such a PE is already under way [17].

Input data

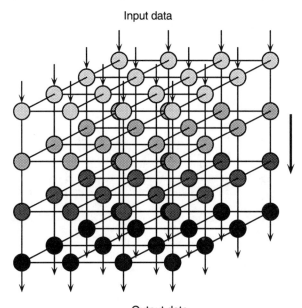

Output data
(Each horizontal two-dimensional plane is connected as a torus.)

Figure 5.12 Three-dimensional cube network.

Finally, bit-processing performance is critical, as we saw with Huffman encoding. Simply increasing VP performance will not speed up the bit stream processing required in the first step. We anticipate that programmable hardware incorporated into each PE will be able to address this performance bottleneck. Suzuki et al. [18] showed that the JPEG-type arithmetic coding function can be efficiently implemented on field-programmable gate arrays (FPGA) and that it is possible to achieve a fully programmable encoding system by using leading-edge logic synthesis with FPGAs. This implies that a future processing element for a programmable DSP system will consist of three types of components: a core processor for communication, a DSP engine, and reconfigurable hardware for bit operations.

5.5 SUMMARY

The performance and architectures required for the digital signal processing o SHD images have been discussed. It was shown that as many as 300 Gop: would be required to achieve real-time SHD video compression. A loosely cou pled parallel architecture was introduced as a promising parallel DSP architec ture that can provide sufficient performance with flexibility. Experimenta results on SHD image processing with the NOVI-II parallel DSP system were de scribed. Based on experiences, techniques needed for the real-time processing of SHD moving images were discussed. The authors believe that advances in VLSI technologies will make it possible to handle SHD still and moving image: as easily as we now handle the present video signals based on the current TV signal format.

References

[1]   Ono, S., and N. Ohta, "Super High Definition Image Communications—A Platform for Medi Integration," *IEICE Trans., Commun.*, Vol. E76-B, No. 6, June 1993, pp. 599–608.

[2]   Sawabe, T., J. Suzuki, and S. Ono, "Performance Analysis of MPEG2-Based Coding Algo rithms for SHD Image Compression," presented at Picture Coding Symp., Sept. 21–23, 1994.

[3]   ITU-T Draft Rec. H.262, "Generic Coding of Moving Pictures and Associated Audio," ISO/IE 13818-2, Nov. 1993.

[4]   Artieri, A., "A Chip Set for Image Compression," *IEEE Trans., Consumer Electronics*, Vol. 3(No. 3, Aug. 1990.

[5]   Lee, E., "Programmable DSP Architectures: Part I," *IEEE ASSP Magazine*, Oct. 1988, pp. 4–19

[6]   Texas Instruments, "TMS320C4x User's Guide," 1991.

[7]   Nakagawa, S., et al., "A 50 ns Video Signal Processor," *ISSCC Digest of Technical Papers*, Fet 1989, pp. 168–169.

[8]   Yamauchi, H., Y. Tashiro, T. Minami, and Y. Suzuki, "Architecture and Implementation of Highly-Parallel Single Chip Video DSP," *IEEE Trans., Circuits and Systems for Video Tect nology*, 1992.

[9]   Sawabe, T., T. Fuii, and S. Ono, "Performance Evaluation of Super High Definition Imag Processing on Parallel DSP Systems," *IEICE Trans., Fundamentals*, Vol. E76-A, No. 8, Au; 1993, pp. 1308–1315.

[10] Intel Corporation, Paragon XP/S Product Overview, 1991.

[11] Inmos, Ltd., *The Transputer Development and iq Systems Data Book*, 1991.

[12] Roscoe, A.W., and C.A.R. Hoare, *The Laws of Occam Programming*, Oxford University Con puting Laboratory, Technical Monograph PRG-53, Oxford, England, 1986.

[13] Logical Systems, "Transputer Toolset," 1992.

[14] LeGall, D., and A. Tabatabai, "Sub-band Coding of Digital Images Using Symmetric Short Ke: nel Filters and Arithmetic Coding Techniques," *ICASSP'88*, 1988, pp. 761–764.

[15] Wallace, K., "The JPEG Still Picture Compression Standard," *Communication of the ACN* Vol. 34, No. 4, April 1991.

[16] Destrochers, G.R., *Principles of Parallel and Multiprocessing*, New York: Intertext Public; tions/McGraw-Hill, 1987.

[17] Fujii, T., and S. Ono, "Interconnection Network Switch Design for Parallel DSP Systems," *I: CAS'92*, May 1992, pp. 1517–1520.

[18] Suzuki, J., F. Colin, and S. Ono, "Arithmetic CODEC From Behavioral Description Based LSI-CAD for Fully Programmable Image Coding System," *Proc. ICASSP'94, Adelaide, Australia, April 1994, pp. II-412–424.*

SHD Image Display Systems 6

6.1 INTRODUCTION

In order to display SHD images without flicker, state-of-the-art display devices are required. At present, CRTs and LCDs are the only devices in the commercial market that cover still and moving images. Near-term display systems for SHD images are also based on CRT and LCD technologies. The requirements for SHD image displays are high frame rate to avoid flicker as well as high spatial resolution This implies that high-speed image memory, which is usually referred to as *frame memory*, is as essential as the display device in SHD display systems.

This chapter discusses the technologies required for displaying SHD images. We divide display devices into two categories: direct-view and projection displays. State-of-the-art direct-view and projection displays are described. The specifications and actual implementation of a frame memory to display SHD moving images are also explained.

6.2 DISPLAY DEVICES

CRTs and LCDs seem to be the only promising display devices for SHD images; direct-view and projection displays are possible. Non-LCD projectors that use space-modulating elements hold great promise, but several technical issues remain to be resolved. Direct-view displays are compact and thus suitable for personal use, whereas projection devices tend to be large. The choice of direct-view and projection displays depends on the application.

The common characteristics of CRTs and LCDs are that they display pixels electronically. Of course, they can display both still and moving images. Displaying a still image just means repeating the same image at the frame frequency rate. As for color, because both types of display depend on fluorescing or reflecting elements, it is difficult to reproduce true black, which can only be obtained by absorbing all light. Rather, these displays use "imaginary black," in which an object appears dark relative to the surrounding bright areas. This is not the same black yielded by printers, who use a pigment in their inks that gives true black. On the other hand, extremely high color reproducibility is obtained because these colors are displayed using light itself.

6.2.1 Direct-View Displays

6.2.1.1 CRT Displays

For a long time the CRT has been the sole method of the direct-view display and much effort has been expended on perfecting it. For this reason, it is sometimes referred to as "the last vacuum tube." However, it has two drawbacks: it requires high voltage and it occupies a fair amount of depth. Color CRTs can be roughly divided into shadow mask and aperture grill (Trinitron) types. The efficiency of electron beam utilization varies greatly between the two, and the difference becomes even more pronounced in the extremely fine tracing of electron beams required for SHD display. In other words, among current CRTs for 2,000 × 2,000 pixel color displays, the Trinitron offers superior performance. For example, a 28-inch Sony monitor can display 2,048 × 2,048 color images at the rate of 60 frames per second [1]. A photograph of this CRT is shown in Figure 6.1 (see color insert). The figure shows the 28-inch CRT displaying an SHD still image. (The original photograph of the image was taken by Shigeru Taguchi.) The monitor has a pixel pitch of 0.31 mm, making it almost impossible to discern the scanned lines with the naked eye. This CRT accepts progressive video signals at the rate of 60 frames per second, with an extremely high white color temperature of 10,000K. What we know today as the CRT was originally designed for air traffic control radar and was heavily promoted by the Federal Aviation Association (FAA). This means that the CRT was not designed for displaying motion pictures. However, the decay characteristics are not particularly long, so normally there are no problems for displaying motion pictures.

6.2.1.2 LCD Displays

LCDs for direct-view displays currently in production cannot display SHD images. The major problem with LCDs is how to make them larger; simply increasing the number of pixels is not thought to be a problem. Currently (1994) display screens of up to 12 inches are being mass produced [2], with advances in maximum size predicted to continue at the rate of a few inches per year. Figure 6.2 shows an LCD for a 12-inch multiscan personal computer display. It produces an extremely clear full-color image, which is probably due to the clarity with which individual pixels are resolved. Not only were personal computers the first application of LCD displays, but they continue to provide most of the demand. However, LCDs will soon be used in 14- to 18-inch displays for workstations; they are likely to completely replace CRTs in this area, due to

Figure 2.2 Example of multiple windows on a display screen.

Figure 2.4 SHD image example (photograph by I. Isshiki).

Figure 3.2a Example of JPEG compression for SHD images: (a) original image; (b) compressed image (20:1); (c) compressed image (48:1).

Figure 3.2b Example of JPEG compression for SHD images: (a) original image; (b) compressed image (20:1); (c) compressed image (48:1).

Figure 3.2c Example of JPEG compression for SHD images: (a) original image; (b) compressed image (20:1); (c) compressed image (48:1).

Figure 6.1 Photograph of a 2,000 × 2,000, 28-inch color display (original photograph of the SHD image was taken by Shigeru Taguchi).

Figure 7.5 Conducting a color-matching experiment.

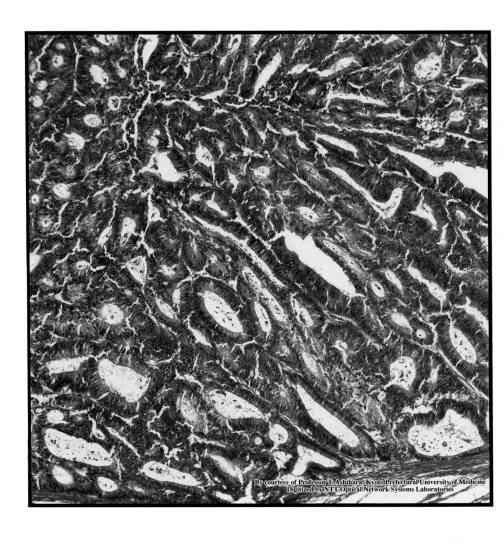

By courtesy of Professor T. Ashihara, Kyoto Prefectural University of Medicine
Digitized by NTT Optical Network Systems Laboratories

Figure 8.2 Example of SHD micrograph image: colonic cancer (*courtesy of* Prof. T. Ashihara, Kyoto Prefectural University of Medicine).

their lower energy consumption, small size, and lack of electromagnetic field (EMF) radiation.

LCDs compare poorly to ordinary CRTs in terms of tonal image quality, but this is gradually improving, and matching the performance of the CRT is potentially possible. When all the criteria for image quality are compared, LCDs beat CRTs in all areas, including color reproduction, except for tonal image quality. The lack of image distortion from the center to the periphery in LCDs is almost impossible to duplicate with a CRT and is a strong argument in favor of LCDs. This lack of distortion means that the LCD can display an image with great clarity.

LCDs can reproduce a wider range of colors because colors are extracted from a white backlight, whereas CRTs combine the spectra of RGB phosphors. Obviously, LCD image quality depends on the quality of the background light and the color filter, but the light source commonly used for the backlight offers more freedom than do the available phosphor spectra and is thus a more ideal source of white. This is easy to understand if we consider that the white light provided by fluorescent lights is better than that created by the phosphors used in CRTs. The technology of the other quality determinant, color filters, is also highly developed, and a wide range of properties is available. Obviously, fluorescent light is not the only means of providing the backlight; in principle, a wide range of sources is available. However, consideration of the shape, heat generation properties, and other factors indicates that fluorescent lights (includ-

Figure 6.2 Photograph of a 12-inch LCD.

ing cold cathode types) or their close equivalents will continue to be used for this purpose.

6.2.2 Projectors

From time to time, groups of people like to view the same large image at the same time, such as at conferences. The only way of displaying images larger than 100 inches for large gatherings such as conferences is to use a projector. Direct-view display devices are impractical for this purpose.

6.2.2.1 CRT or LCD

Some of the projectors use CRTs, while others use LCDs. Because CRTs used for projectors must have extremely high luminance, phosphors different than those found in direct-view CRTs are used [3]. The only other differences from ordinary CRTs relate to the electron beam, which is much stronger, and because the unit is only several inches in diameter, it does not require a very hefty deflection system for the beam. The white color projected by some commercial versions onto a screen several meters away can reach 2,500 peak lumens. The luminance of projectors that use LCDs depends on the light source rather than the LCDs. The determining factors are the heat, lifetime, and wavelength distribution of the lamp used as the light source and the deterioration of the LCDs in response to the heat. Since LCDs employ changes at the molecular level, they tend to age rapidly in response to heat.

The fact that attempts are underway to develop projectors that use space modulation elements other than high-luminance LCDs indicates that researchers see no possibility of solving the problem of the heat-dependent aging of LCDs. Yet, perhaps this despair is premature, since research into LCD materials has just begun and there are many other unexplored technical areas. One approach to the inherent problem of heat aging is to find some method of cooling the LCDs, but this approach is unlikely to generate much interest, since it would sacrifice the compact size that is the major advantage of LCDs.

6.2.2.2 Type of Projection

Projectors can project the image onto the screen either from the front or from the rear of the screen. In front projection, images are projected onto a reflective screen from the front. In rear projection, images are projected from the rear onto a translucent screen. Rear-projection devices often insert reflecting mirrors between the projector and the screen in order to shorten the light path. The compactness of the rear-projection method is frequently the decisive factor in its favor. However, the image on translucent screens is unavoidably dimmer than that projected onto reflecting screens.

Conventional high-gain screens (i.e., having a high reflection coefficient) are coated with fine glass powder, but the improved luminance of recent projectors permits ordinary white cloth to be used. This prevents glare from the screen, which in turn reduces eye fatigue from long-term viewing.

6.2.2.3 Lens System

The lens system may use a single bulb or three to six bulbs. Multiple lens bulbs are employed to minimize light loss in the optical system, but this makes it hard to coordinate the focal points of all the lenses, with the unavoidable result that the image is dim and the system takes up more space. It is also difficult to adapt the focus to the installation site, making this system impossibly unwieldy for applications that require portability. Six lens bulbs achieve very high luminance levels, but adjusting the focus becomes even harder and the size of the overall system even larger.

Generally speaking, most projectors that use LCDs are designed for compactness and for the same reason employ only a single bulb. As a result, the performance of these projectors is unimpressive. In contrast, projectors that use CRTs are designed for high performance; thus most use three lens bulbs and many have electronics that can handle a variety of signals for multiscanning.

Barco is manufacturing a high-resolution projector that uses CRTs. The resolution of this projector does not compare to the CRT shown in Figure 6.1, but at present it offers the highest performance and functionality, where "high performance" is understood as the ability to display almost all video signals regardless of whether they are interlaced or noninterlaced.

6.3 FRAME MEMORY SYSTEMS

6.3.1 Overview

Frame memory refers to systems for storing digital image data for several frames or several hundred frames and displaying these images as motion or still pictures. Accordingly, these systems comprise a memory module for storing the image data, a module built around D/A converters for generating image signals from the digital data stored in the memory module, an interface module for loading data into the memory module, and a control module for controlling the entire system.

Normally, the memory module, which must provide high-speed access, is based on dynamic RAM (DRAM). The capacity of these memory modules depends on whether the object is to display still or motion pictures. For still pictures, the ability to hold data for two frames in a pair of buffers is adequate, but a vast amount of memory is needed for motion pictures, since data must be held for the minimum number of frames required for display. Dedicated memory for operating the read-in and read-out ports independently has been developed and is widely used for image display applications. The demand for such memory is expected to keep pace with the increasing use of graphical user interfaces (GUI).

Three types of interface modules are used for frame memory systems: (1) those that connect to the computer's internal bus, for example, VME and S-BUS; (2) those that connect to the input/output (I/O) ports for external disks, for example, small computer system interface (SCSI) and Fiberchannel; and (3) those that connect to ports for local-area networks (LAN), for example, Ethernet, FDDI, and asynchronous transfer mode (ATM) LANs. Most frame memory systems for still images use the first type, since data transmission requirements are modest enough that the internal bus is occupied only for short periods of time, and the effect on other computers on the network is minimal.

The module used to generate video signals from the digital image data centers on three D/A converters, one each for R, G, and B. The quality of the displayed image depends on the properties of these converters. Obviously, the best quality requires converters that can handle long words, permitting the fastest possible operations. Currently available D/A converters for high-speed image processing operate at 8 bits with 1-GHz sampling.

For frame memory systems that connect the computer's internal bus to a disk I/O port, a new program module, called a *device driver*, must be added to the operating system. The device driver is necessarily dependent on the model of computer and type of operating system. On the other hand, frame memory systems that are connected to a LAN port do not suffer from this inconvenience and can be connected to any type of computer relatively easily. However, in this case the interface module will require an extra computer to be added to the frame memory system just for this purpose, thus raising the cost of the entire system.

6.3.2 Frame Memory Architecture for SHD Images

The problems of frame memory for SHD images arise from the requirement that as the number of pixels increases, vast amounts of image data must be processed rapidly. For motion pictures, three D/A converters must supply data at a rate of about 6 Gbps (with each D/A converter operating at about 2 Gbps). The central architectural problem is how this requirement can be met.

For any type of storage medium, the access time generally varies in inverse proportion to the cost. In other words, storage media offering high-speed access tend to cost more than their low-speed counterparts. In terms of access rate and cost per bit, storage media can be ranked as follows: floppy disks, magneto-optical (MO) disks, hard disks, DRAM, and static RAM (SRAM). Note that the requirement for random access eliminates magnetic tape from consideration. The ranking given here is based on the number of bits that can be recorded per unit volume.

The above ranking system does not take into account the fact that access time in DRAM and SRAM does not depend on the address, but this does matter with the three types of disk. Since the maximum disk access time must be calculated for applications that demand a constant data rate, access time must be an order of magnitude greater for disks than for DRAM and SRAM.

Only one method exists for the high-speed access of data on slow-speed storage media, and that is to create a broad-band data bus by arranging several output ports for several low-speed storage devices and connecting this bus to a parallel/serial converter. Obviously, this assumes that the low-speed storage devices all work at the same speed and are synchronized.

In the first SHD frame memory system, we have developed jointly with Mitsubishi Electric, we have used 4-Mb DRAMs for image data storage, with a broad-band bus to the 4-Mb DRAM storage [4]. This architecture is illustrated in Figure 6.3. The major technical problem was how to make the broad-band bus work at high speeds. More specifically, we had to minimize the signal delay caused by the long circuit path necessitated by using so many DRAMs. This was done by optimizing the design of the printed circuit board to shorten path lengths and by judicious selection of the connectors and design of the backplane. We also used an ECL gate array in order to speed up the circuits around the D/A converters.

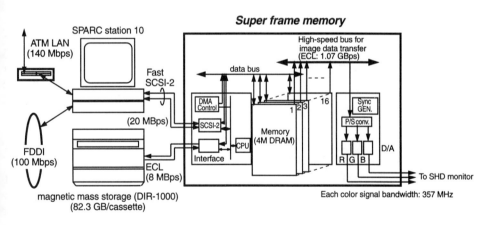

Figure 6.3 Frame memory architecture for SHD images.

6.3.3 Specifications of the Developed Frame Memory

6.3.3.1 The First Generation

Table 6.1 shows the frame memory specifications for the first system we developed (see Figure 6.4). This system can store 240 frames using about 8,000 4-MB DRAMs. This is equivalent to 4 seconds of SHD motion pictures at 60 frames per second. This system was designed so that the 4-MB DRAMs can be replaced with 16-MB DRAMs and so achieve 16 seconds of motion picture display.

Figure 6.4 Photograph of an SHD frame memory (early version).

Table 6.1
Specification of SHD Frame Memory (Early Version)

Item	Specifications
Video output (RGB)	0.714Vp-p, 50Ω
External sync (HD, VD)	TTL (negative), 75Ω
Spatial resolution	2,048 × 2,048 pixels
Temporal resolution	Noninterlaced 60 frames per sec
Sampling frequency	357 MHz, Square sampling
Number of frames	240 frames (expandable up to 960 frames)
Display period	4 seconds (up to 16 seconds)
Address of frame memory	2,048 (V) × 2,560 (H) × 3 (RGB) × 240 (up to 960) frames
Display functions	Slow, still, reverse play, zooming and scroll
Power consumption	3 kilovoltampere (kVA)
Size	920 (W) × 720 (D) × 1,300 (H) mm
Weight	280 kg

Four seconds is perhaps too short for adequate subjective evaluation of image quality, but it is sufficient for various other image quality evaluations of the motion picture. Moreover, 240 still images form an extremely effective presentation. Therefore, this system is quite useful when combined with other elements in a multimedia presentation.

Figure 6.5 shows the memory board; Figure 6.6 shows the board on which the D/A converters are mounted. This system is connected to two SCSI ports for high-speed data transfer with the host computer.

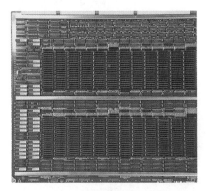

Figure 6.5 Photograph of a memory board used in an SHD frame memory.

Figure 6.6 Photograph of a D/A converter board.

The frame memory system is connected to five coaxial cables (R, G, B, VSYNC, HSYNC). Normally, the cables can extend for up to 10m without noticeable image deterioration due to delay.

6.3.3.2 The Second Generation

Figure 6.7 shows a second-generation system developed to increase the practicability of SHD imaging systems. The basic features are based on the previous system, with the main differences being the use of 16-MB DRAMs to reduce size. The range of compatible host computers is broadened by using FDDI and

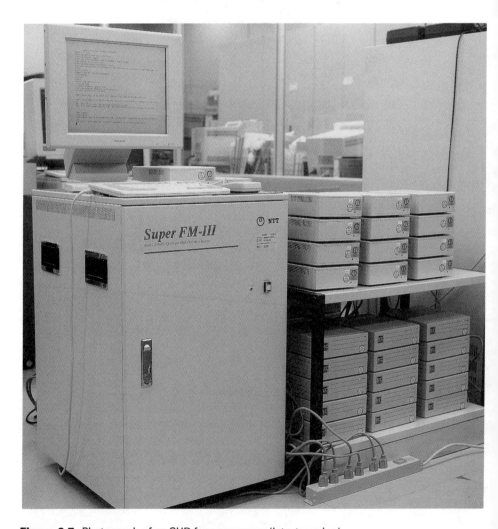

Figure 6.7 Photograph of an SHD frame memory (latest version).

ATM-LAN for the interface. Obviously, this can also be connected to the public communications network via a router. The specifications of this system are shown in Table 6.2.

Table 6.2
Specification of SHD Frame Memory (Latest Version)

Item	Specifications
Video output (RGB)	0.714Vp-p, 50Ω
External sync (HD	TTL Negative, 75Ω
Spatial resolution	2,048 × 2,048 pixels
Temporal resolution	Noninterlaced 60 frames per sec
Sampling frequency	357 MHz, Square sampling
Number of frames	256 frames
Display period	4.26 seconds
Address of frame memory	2,048 (V) × 2,560 (H) × 3 (RGB) × 256 frames
Display functions	Slow, still, reverse, zooming and scroll
Power consumption	700 volt ampere (VA)
Size	800 (W) × 480 (D) × 600 (H) mm
Weight	95 kg

The experience of developing this system demonstrated the feasibility of putting all frame memory on a single board. More accurately, if daughter boards are used, 64-MB DRAMs can be squeezed onto one board according to how the DRAM is packaged. If 256-MB DRAMs are used, it will soon be possible to offer this type of frame memory system as an add-on board for an ordinary computer.

6.3.4 Towards Future Multimedia Systems

The above frame memory systems for SHD images hold the possibility of future multimedia applications [5,6]. Furthermore, the authors believe that our systems have demonstrated the possibility of handling high-quality motion pictures in a practical multimedia system. It is expected that general trends in DRAM development will support the needs of image display applications. However, only future developments will show whether this kind of multimedia system will be called a dedicated personal computer or workstation, or a motion picture editing device. The latest trend can be observed in the motion picture editing devices. An example of such devices is manufactured by AVID Corporation (see Figure 4.1). The only hardware requirements of this system are a JPEG board added to a commercially available Macintosh Quadra with a high-speed magnetic disk. This system can be used for extremely rapid and easily implemented editing of NTSC motion pictures.

Some sort of motion picture editing capability is likely to be a standard feature on future multimedia systems. This is due to the extreme redundancy of most unedited motion pictures. The unedited version can erode the value of even a motion picture produced with great care. Conversely, this situation creates conditions that greatly foster the emergence of a new operating system or GUI. But at present it is extremely difficult to declare what form such an operating system or GUI will take. The most sensible advice on this topic has been offered by Alan Kay, who said, "The best way of predicting the future is to invent it."

Nowhere is the meaning of this proverb more evident than in the practical applications made possible by the combination of frame memory systems with a personal computer or workstation and an ultrahigh-speed digital communications network (150 to 600 Mbps). After all, Alan Kay does not speak as an academic futurologist, but as the man who developed the Xerox Alto system [7]. In other words, he is the wizard who personally created the foundations for present-day multimedia. It is quite clear that without the ALTO system to lay the groundwork, the Macintosh would never have been developed to its present form at the present time. In this sense, our SHD frame memory system may be as valuable as the ALTO system.

6.4 SUMMARY

We have discussed the technologies required for displaying SHD images. We divided display devices into two categories; direct-view and projection displays. The state of the art of direct-view and projection displays were described. The specifications and actual implementation of a frame memory to display SHD moving images and future trends were discussed.

References

[1] Sony Corp., Technical Catalog of DDM-2802, 1992.
[2] Tannas, L.E., "Evolution of Flat-Panel Displays," *Proc. IEEE*, Vol. 82, No. 4, April 1994.
[3] Gorog, I., "Displays for HDTV: Direct-View CRTs and Projection Systems," *Proc. IEEE*, Vol. 82, No. 4, April 1994.
[4] Murakami, T., H. Ohira, O. Tanno, R. Suzuki, M. Wada, T. Saito, K. Ogura, and K. Asai, "The Development of Super High Definition Image Storage System," *Proc. ICASSP'92*, 1992.
[5] Inoue, K., M. Tomura, S. Nonaka, S. Kuga, G.D. Alexander, and T. Nakamura, "Study for Developing Applications of Super High Definition Images," *Conf. on High Definition Video*, SPIE Vol. 1976, Berlin, 1993, pp. 202–209.
[6] Ono, S., N. Ohta, and T. Fujii, "Media Integration Platform on Super High Definition Images—Parallel Digital Signal Processor Approach," *Conf. on High Definition Video*, SPIE Vol. 1976, Berlin, 1993, pp. 224–235.
[7] Kay, A., and A. Goldberg, "Personal Dynamic Media," *IEEE Computer*, March 1977, pp. 31–41.

SHD Imaging Experimental System

7

7.1 INTRODUCTION

As shown in the previous chapters, SHD images provide a qualitative advance in image quality, and their implementation will necessitate a number of technological advances. These advances are required in diverse fields: input, display, storage, and signal-processing devices. In this chapter, we introduce an actual experimental system developed in the authors' laboratories and describe the equipment employed in this system. The experiments and the investigations that have been carried out with it are also explained.

7.2 EXPERIMENTAL SYSTEM EQUIPMENT AND PERFORMANCE

7.2.1 Overview

The most fundamental difference of this experimental system from older technologies is the larger number of pixels. In addition, the extremely broad range of applications in which SHD images will be used and the very high quality demands mean that engineers will have to pay close consideration to many details of image processing that could formerly be ignored. Because there is so much new ground to be covered, experimental systems will have to include input, processing, storage, communication, display, and hard copy of SHD images.

This is the concept that motivated the authors' construction, beginning in 1989, of the experimental system illustrated in Figure 7.1. All of the equipment is controlled by workstations via Ethernet or FDDI [1]. All data communication occurs via the network. The specifications of the overall system are provided in Table 7.1.

We have used this system to carry out the following investigations.

1. Evaluation of compression algorithms;
2. Evaluation of parallel processing algorithms and systems;

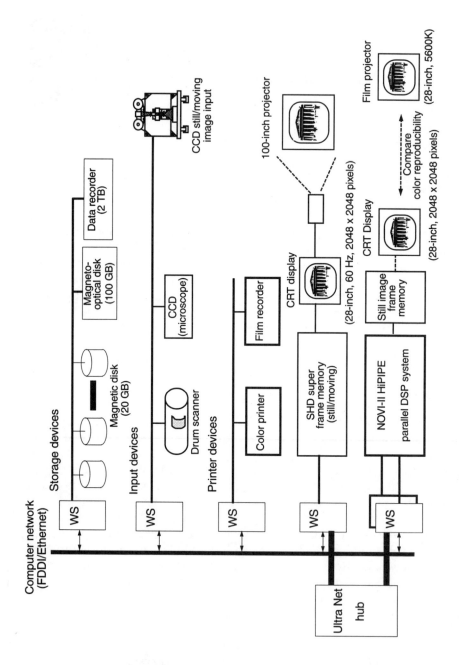

Figure 7.1 Overview of experimental SHD image system.

3. Application feasibility studies;
4. Experiments with communication networks and interfaces;
5. Creation, maintenance, and distribution of standard reference images;
6. Assembly of fundamental data for evaluation of SHD image quality (e.g., characteristics of input/output devices).

Table 7.1
Experimental SHD Image Input System

Input Equipment for Digital Still and Moving Images

Movie camera	
Film size	70 mm (type 1 film)
Aperture size	57 × 57 mm
Frame speed	variable, 6–125 fps
Film input equipment	
(automatic film advance)	4096 × 4096 pixels (CCD line sensor)
Pixel resolution	12 bits/pixel
Quantization bits	RGB
Color space	halogen lamp (3200)
Light source	5000 K, 5800 K, 6500 K
Light source filters	
Drum scanner	
Image sensor	Photomultiplier
Aperture size	6.25–1600 mm
Quantization bits	12 bits/pixel
Color space	RGB
Micrography input	
Pixel resolution	4096 × 4096 pixels (CCD line sensor)
Quantization bits	14 bits/pixel
Color space	RGB

Signal-Processing System

Performance	120 Mflops × 128 processing elements
Accuracy	32 bit floating point
Memory	

Frame Memory

Resolution	2048 × 2048 pixels
Depth	8 bits/pixel
Frames	240 frames
Frame rate	0–60fps

Table 7.1 (continued)

Display System	
CRT	28 inch, 8 bits/pixel × RGB
Video projector	100 inch, 8 bits/pixel × RGB

Printer	
Photoprinter	3800 × 2500 pixels, 8 bits/pixel × RGB
Film printer	8192 × 8192 pixels, 8 bits/pixel × RGB

Computers	
Network	ATM LAN, FDDI, Ethernet, Ultranet
Magnetic disk	Approx. 20 GB
Magneto-optical disk	2 TB

7.2.2 Input of Still and Moving SHD Images

The organization of the SHD image input system is shown in Figure 7.2. Still images are input by using a scanner to digitize film exposed with a medium- or large-scale still camera, or by using a CCD digital camera to directly capture a digital image. As of 1994, no SHD video camera had yet been developed; moving SHD images are input by digitizing individual frames taken with a high-speed 70-mm movie camera. As discussed in Chapter 4, the authors have developed a custom film digitizer for this purpose. Below we briefly outline the performance and our method of applying each of these devices. Let us emphasize that achieving the SHD image quality levels we are aiming for takes great care and attention to many details. It does not suffice to simply purchase commercial products and connect them to a network; a modicum of empirically derived know-how is required.

1. *CCD line scanner head.* Currently, CCD line scanners with 2,000 × 2,000 pixel resolution are commercially available. At the start of our experiments, we used an Eikonix model 1412 CCD line scanner to input still images from photographic paper. It soon became clear that achieving resolutions of 2,000 × 2,000 pixels or better with just this system was problematic. To obtain sufficient dynamic range, we found that directly digitizing color reversal film was preferable to working with photographic paper. Further, we discovered that noise found its way into the pixels from small building vibrations and other causes. We circumvented this problem by mounting the entire apparatus on a pneumatic antivibration table. (Fortunately we did not have any earthquakes during the digi-

tization.) The intensity of the lamp was increased to match the CCD sensitivity, and the camera head was mounted on a precision stage controlled by an automatic focusing device. We took advantage of the library provided with the scanner to develop custom control software that runs on a Sun workstation. It is possible to produce 4,000 × 4,000 color scans with this scanner head in about 5 to 15 minutes. Using a resolution test chart, we have verified that the actual resolution achieved by this system is about 2,900 pixels across.

2. *Drum scanner.* A drum scanner is used for applications requiring more than 4,000 × 4,000 pixels, or when the film to be scanned is larger than 4 × 5 inches. We use an Optronics ColorGetter II, which can handle film sizes up to 8 × 10 inches and delivers a maximum of about 32,000 × 40,000 pixels for 8- × 10-inch originals (Figure 4.4). As with the CCD scanner, we developed custom control software that runs on a workstation and acquires 12 bits (of intensity, not density). It takes about 15 to 30 minutes to input a 4,000 × 4,000 color image on this system.

3. *70-mm high-speed movie film camera.* The aperture size of common 35-mm movie camera film is only 24 × 18 mm, and 70-mm movie film is approximately 70 mm across, but only half that high. It is difficult to acquire 2,000 × 2,000 images with this format, so we use a Photosonic 70-mm high-speed movie film camera (70-mm-10R). This camera was designed for instrumentation applications. It is used, for instance, by NASA. The aperture size of this camera is 57 × 57 mm, and the resolution is adequate for capturing SHD images. We used this camera, shown in Figure 4.2, to capture moving test images at 60 frames per second. These films were digitized frame by frame to produce sample SHD movies.

4. *70-mm film digitizer* [2,3]. Moving SHD images were input by digitizing individual frames taken with the above high-speed 70-mm movie camera. A problem that should be resolved is to reduce the time taken by frame-by-frame digitization. We have created a prototype of a semiautomatic 70-mm movie film digitizer by combining a precision film advance unit with the line scanner discussed above. A photograph of the system appears in Figure 4.5. Precise film advance is ensured by a mechanism with registration pins designed for film measurement. It takes about 5 minutes to get a full-color 2,000 × 2,000 image from each 70-mm movie film frame.

5. *Input of microscopy images.* There is strong demand for the digitization of microscopy images in medicine, particularly in pathology. In order to investigate the issues, we prototyped a device that can input digital images directly from a microscope and called in professional pathologists to evaluate it with us. The system uses a Leaf Digital Camera by Scitex, which outputs 2,000 × 2,000 pixels with 8 bits on each of the R, G, and B channels. A photograph of the system is shown in Figure 7.3. It takes about 90 seconds to get a 2,000 × 2,000 color microscopy image.

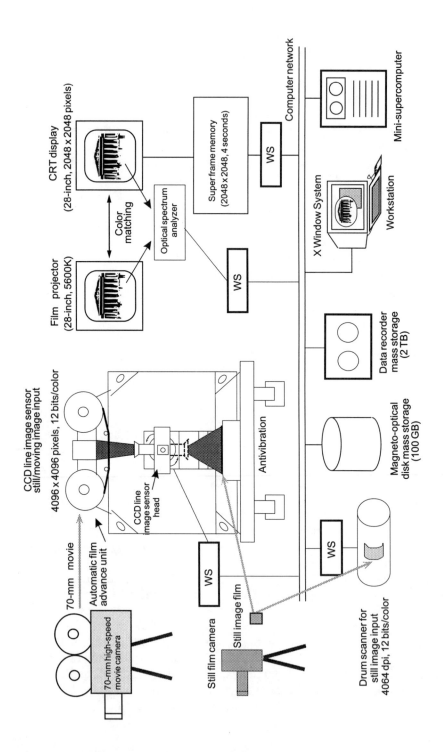

Figure 7.2 SHD image input system.

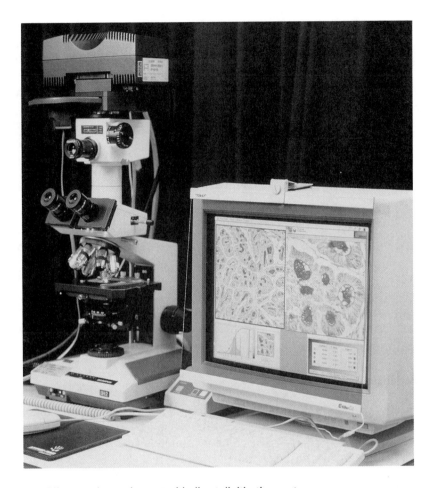

Figure 7.3 Micrography equipment with direct digitization system.

7.2.3 Display

For experimental purposes, it is better to prepare many kinds of displays in an experimental system. However, strictly speaking, there are no devices other than CRT monitors that can display more than 2,048 × 2,048 color pixels with sufficient quality. Here, "sufficient quality" means that there is virtually no difference in terms of quality between the original and displayed images in most of applications. For quality, flicker-free is an important factor as well as spatial resolution, color reproducibility, and brightness.

We displayed SHD images on a Sony DDM2802, a 2,048 × 2,048 color monitor originally developed for air traffic control stations. It supports a frame rate of 60 frames per second. We also evaluated other CRT monitors from other companies that were stated to have 2,000 × 2,000 resolution. We also evaluated their quality and chose the Sony monitor due to its high resolution and precise color reproducibility. Our experimental system included a 100-inch projector, which is effective when an audience of any size is to view SHD images on a large screen. The projector is a Barco Graphics 1200, with a three-color light source based on three bulbs. Although it does not deliver 2,000 × 2,000 resolution, it is useful for viewing SHD images.

We needed a high-speed frame memory to display 60-frame-per-second SHD images. The frame memory was developed by Mitsubishi to our custom specifications. The details are given in Chapter 6 of this book.

7.2.4 Signal-Processing System

For SHD image experiments, a high-speed I/O channel was attached to our NOVI-II parallel signal processing system. The details of the resulting system, called a highly parallel image processing engine (HiPIPE), are given in Chapter 5. In addition to HiPIPE, our experimental system incorporates a number of mini-super computers, including Convex and Challenge, connected via the high-speed Ultra Net and other networks. These computers allow us to check out compression algorithms.

7.2.5 Hard-Copy Output

In order to evaluate input (digitizing) characteristics of the experimental system and to carefully examine the degradation caused by compression, we needed printers or film recorders. As a color printer, we use the Fujix Pictrography 3000, a laser exposure thermal transfer printer that delivers 3,800 × 2,759 full-color pixels at A4 size [4] (see Figure 7.4). The resolution and color reproducibility of the output is similar to that of a typical photograph. Although the printer could benefit by further adjustment, its stability and reproducibility are good enough for our experimental purposes. In order to get films, we use the Solitaire color film recorder, which offers a maximum resolution of 8,192 × 8,192.

7.2.6 Network

The general network that connects the computers in our laboratory is separate from the network that supports the experimental system. All data can be transferred between devices very simply via workstations. The network includes Ethernet, FDDI, and Ultra Net components. Ethernet, which has a maximum speed of 10 Mbps via coax cables, is insufficient for SHD data transfer. So we installed FDDI interfaces in the key workstations to control the frame memory,

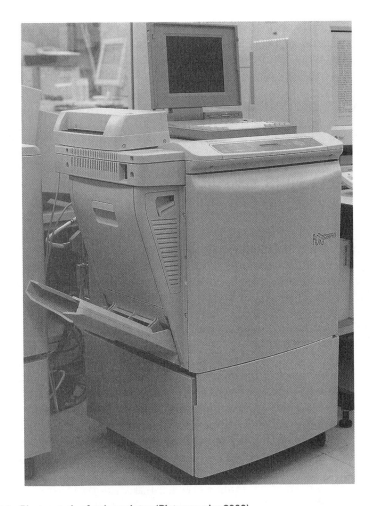

Figure 7.4 Photograph of color printer (Pictrography 3000).

NOVI-II HiPIPE and other devices for achieving high-speed data transfer. The FDDI provides at most 100-Mbps data transfer via optical fibers. The actual data rate achieved between workstations, however, was about 30 Mbps. Ultra Net with a rate of several hundred Mbps is used only for high-speed transfer between the frame memory and NOVI-II. For the transmission of SHD image data to other locations, ATM interfaces offering 150 Mbps will be needed [5].

7.2.7 Image Data Storage Devices

Storage of SHD images is supported by gigabyte-class disks on each workstation, a 100-GB magneto-optical mass storage system, and a 2-TB data tape recorder mass storage system. Frequently used data are stored in a workstation's disk and transferred to HiPIPE when necessary. Other SHD data are usually stored in the magneto-optical mass storage system.

7.3 APPLICATION INVESTIGATIONS

1. *Creation of test images for reference.* As explained in Section 2.5, the original images were produced by professional photographers using 4 × 5-inch or Brownie format film, with careful attention to optimal lighting conditions, focus, and shutter speed. This film was then converted to 4,000 × 4,000, 12-bit/pixel data using the CCD line sensor head or drum scanner systems described above. The data were then shading- and gamma-corrected to produce the test images. We consciously tried to produce test images that included many types of backgrounds, colors, materials, and textures. These images are available to other investigators who wish to evaluate SHD images (see Figure 2.4 in Chapter 2). Four of the test images were distributed to participants at the 1994 Picture Coding Symposium (PCS'94). The same images were used to evaluate image algorithms in a paper also presented by the authors at PCS'94 [6].

2. *Input of 70-mm instrumentation movie film.* At the current time, we have a small collection of SHD movies of scenery, people, and the like, digitized with our custom film digitizer. We have verified that the spatial resolution and motion reproducibility of the system is adequate for the production of SHD films. However, because the resolution is so high, films shot using the usual film and video techniques tend to have an unnatural appearance. It is clear that new shooting techniques will have to be developed for this medium.

3. *Evaluation of medical images.* We carefully digitized x-rays and photomicrographs of cells provided by physicians and pathologists. Physicians evaluated the resulting 2,000 × 2,000 SHD images on a CRT monitor. Our informants confirmed that these images were of sufficient quality to permit diagnosis, within certain limits. In particular, pathologists evaluating cell micrographs indicated an appreciation for the high color reproducibility; this contrasts with the color signal compression of HDTV, which can lead to incorrect diagnoses in that medium. The pathologists also appreciated the near absence of flicker in our system, which makes it easy to measure small cell features and to observe the screen for extended periods of time without fatigue. See Chapter 8 (Section 8.2) for a further discussion of medical applications.

4. *Kimono ordering system.* Japanese kimonos are often constructed of fabric of such high quality that it might be termed a work of art in itself. It would be nice if the dealer could meet in detail each customer's

needs, even when a particular item is not in stock. We built a prototype system to determine whether or not SHD images could be used to accurately determine the customer's tastes. The result was affirmative. The system's concept is described in more detail in Chapter 8 (Section 8.5).

5. *Color-matching experiment.* We have proposed a technique, called *color matching*, for determining the color reproducibility of input digital images, and have performed experiments to evaluate the technique [7]. In order to achieve good color reproducibility, we applied corrections to actual digitized images and investigated the results. Figure 7.5 (see color insert) shows the color of digitized images displayed on a CRT being compared to the color of the original photograph, recorded on film. The result is that the sum of squares color error (the difference in color space) can be reduced by 30%. This produces a subjectively significant improvement. A closer investigation of the appropriate color space and the basis for optimization will be necessary in order to achieve more improvement. Details of the color-matching technique are given in Appendix B.

7.4 SUMMARY

This chapter described an SHD imaging experimental system in the authors' laboratories and some experiments that have been conducted using the system. The system is not yet completed and needs more development and testing to reach the quality demanded by a variety of SHD imaging applications. This is because the applications will range from technical to cultural fields, from medicine to arts. A detailed discussion on future applications is given in Chapter 8.

References

1] Mazzaferro, J.F., and A.A. Dell'Acqua, "FDDI Technology Report," Computer Technology Research Corp., 1992.

2] Furukawa, I., K. Kashiwabuchi, and S. Ono, "Super High Definition Image Digitizing System," *Proc. ICASSP '92*, Vol. III, March 1992, pp. 529–532.

3] Fujii, T., I. Furukawa, and S. Ono, "Motion Video Sequence Processing System for Super High Definition Images," *Proc. Society for Information Display 1992 International Symp.*, May 1992, pp. 310–313.

4] Kubo, M., et al., "A Color Hardcopy System With Improved Color Fidelity," SPIE Vol. 1670, *Color Hard Copy and Graphic Arts*, 1992, pp. 478–485.

5] Partridge, C., *Gigabit Networking*, Reading, MA: Addison-Wesley, 1993.

6] *Proceedings of PCS '94*, Sacramento, CA, Sept. 21–23, 1994.

7] Kashiwabuchi, K., I. Furukawa, and S. Ono, "Film Based Motion Picture Digitizing System for Super High Definition Images," *Proc IS&T/SPIE Symp. on Electronic Imaging Science & Technology '94*, Vol. 2173-18, Feb. 1994.

Applications of SHD Imaging 8

8.1 INTRODUCTION

Human intellectual activity may be rational or emotional, but in either case it relies heavily on visual information (i.e., image data). SHD image data provide a level of quality acceptable to the most refined intellect and sensibilities and is thus suitable for application to a wide range of cultural and scientific activities. At present, it is too early to predict the entire range of applications, just as in the past it was impossible to predict all the uses to which the zipper or any other invention might be put before it had been invented.

In this chapter, we will discuss the shape of future applications of SHD imaging, focusing on medicine, art, printing, and education. Indeed, the range of possible applications is so wide that at present it is difficult to organize all the material. In this sense, the statements made in this chapter are woefully incomplete, representing perhaps merely a fraction of the full panoply of applications the future holds.

8.2 MEDICAL APPLICATIONS

8.2.1 The Role of Imaging in Medical Diagnosis

Data used in medical diagnosis are frequently obtained in the form of image data. The data may be obtained directly from visible light, or some method may be used to convert the data into visible images. The principle of this may be explained as follows. The human body is completely impenetrable to visible light. However, other electromagnetic waves and sound waves can penetrate or be reflected in the human body, albeit in attenuated form, making it possible to derive reasonably clear assumptions about what is happening inside the body. In other words, modern medical diagnosis has become heavily reliant on presenting data in visual form. This fact enables us to understand why it is almost impossible for a blind person to conduct a medical diagnosis.

The fact that few images are obtained for medical diagnosis directly from visible light is due to constraints on the observation of data from the body that can be made relatively easily. Here, observations that can be made "relatively easily" means "noninvasive," such as endoscopic examinations. Obviously "relatively easily" does not mean "without discomfort to the patient." Normally, an endoscopic examination also involves a biopsy, where a sample of tissue is taken from an organ for study under a microscope.

There are many cases when noninvasive means alone do not provide sufficient data for diagnosis. Therefore, other noninvasive methods have been developed, such as the x-ray. The x-ray has a long history and has become the most common form of medical imaging. X-ray imaging devices use an x-ray source (i.e., a signal source) to expose an image on x-ray film. The signal sources with the next longest history are the radioactive isotopes in scintigraphy, which also uses film as the imaging medium.

Medical diagnostic imaging devices that use sound waves typically use ultrasound, because the attenuation of ultrasound in the human body is minimal and the reflected wave provides a strong signal. Since ultrasound has a negligible effect on the body and can produce moving pictures, it is used for diagnosing embryos. However, the resolution of such images is poor.

Recently, x-ray CT scanners, MRI, PET, and other devices for measuring conditions inside the body have become widespread, and the market penetration of CT scanners and MRI is particularly noteworthy [1]. Massive calculations are performed to synthesize cross-sectional images of the human body. Recent computing power has advanced to the point where even three-dimensional images can be synthesized. Only the most powerful modern computers can perform the necessary number of calculations fast enough to synthesize images in a relatively short time.

Since these diagnostic tools use sensors to detect the penetration or reflection of electromagnetic or ultrasound waves, the data thus obtained can be optimized by sophisticated algorithms. Obviously, the relationship between the signal source and the sensors must be precisely controlled in order to obtain the best possible measurements. The complexity of these machines is such that even a summary description may be safely omitted from a discussion of SHD imaging.

8.2.2 Characteristics of Medical Imaging

The characteristics of the images obtained from various diagnostic devices may be evaluated by three criteria: (1) pixel count, (2) accuracy, and (3) color reproducibility. The *pixel count* refers to spatial resolution; simply, the more pixels the larger size film or CRT is required for display. *Accuracy* refers to the amount of digital data required to represent each pixel (the word length, including that of the luminosity signal). As is well known, the sensitivity of human vi tion is far from linear, providing an extremely broad dynamic range and high adaptability. Nonlinearity means that those gradations that can be discriminated depend heavily on the value of luminance signals. There are two types of digital signal representation: floating-point representation, with separate exponent and mantissa, and fixed-point representation. In consideration of the inherent nonlinearity of the human visual system, the floating point representation is best suited for evaluating dynamic range. However, since most A/D converters in actual use rely on fixed-point arithmetic, this is what we will use here. Nonlinearity refers to the fact that the ability to distinguish gradations is significantly greater for extremely dark than for extremely bright images.

The last criterion, that of color reproducibility, refers to the number of distinct colors that can be displayed. Strictly speaking, the number of displayable colors is not exactly identical to color reproducibility. However, there is a certain proportional relationship between color reproducibility and the number of displayable colors. For the purposes of this discussion we will assume that color reproducibility represents the number of displayable colors. This proportional relationship can be readily understood if we consider the region shaped like a yacht sail in the CIE standard colorimetric system diagram or three-dimensional models of color space. It has been proven that this sort of representation is extremely effective.

The graph in Figure 8.1 evaluates the images produced by medical diagnostic devices using these three criteria. In this figure, the performance of the optical system and film is considered. (See Chapter 2 for film resolution.) Obviously, the performance of diagnostic imaging devices varies considerably. The graph rates well-known or widely used machines and is not intended to be used as an exhaustive critical inventory of performance.

8.2.3 Medical Applications of SHD

Figure 8.1 demonstrates that SHD images of the most advanced technology, with 2,048 × 2,048 pixels, a palette of 16 million colors, and 256 gradations (8 bits of luminosity), are sufficient for supporting nearly all existing types of medical images. Obviously, some types of images require higher quality, but future developments in technology will make up for this and bring them within the purview of SHD imaging. Moreover, evaluations by a number of representative medical specialists have confirmed that those applications that cannot yet be addressed by SHD imaging are not significant enough to prevent its practical application.

This endorsement of currently available SHD imaging means that all the visual data required for medical diagnosis can now be manipulated on a computer, providing paperless, filmless patient records that can be transferred over LAN and B-ISDN anywhere they are needed. Physicians can now consult such

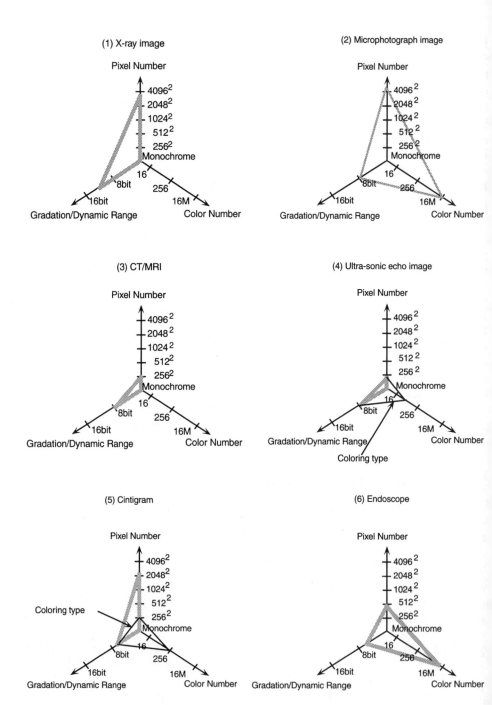

Figure 8.1 Characteristics of medical images. (The optical system performance is considered.)

files and obtain reasonably accurate data on a patient's history, including previous surgery, drug administration, and the results of various tests. The patient thus receives a more accurate diagnosis from the physician, risky drugs are avoided, and tests are not repeated unless necessary. On-line availability of patient files also makes it easier for general practitioners to work effectively with specialists to establish an accurate diagnosis. There is no need to add that these are very desirable goals from a social standpoint as well. Obviously, steps must be taken to ensure that the patient's privacy is maintained, but this can be easily done through modern encryption technology, including the use of digital signatures. The practical barriers to full implementation of such a system are thus legal, organizational, and social, since the issues of quality and cost have already been solved by technology.

An example of microphotograph images is shown in Figure 8.2 (see color insert). Medical experts have confirmed that the quality of digitized SHD images of this type is adequate for diagnostic purposes. (See Chapter 7 for the microscopy system used to take the images shown in Figure 8.2 for pathological studies.) It is clear that SHD imaging will encourage the use of electronic media in medicine. However, any deterioration of the quality of medical image data is to be avoided, since it may result in misdiagnosis. Therefore, medical image compression will be allowed only in certain cases, such as medical emergencies. Otherwise, troublesome issues of liability under civil and criminal law may occur. It is thus anticipated that lossless encoding (i.e., fully reversible encoding) will be widely used in the field. The compression rate for lossless encoding is low (never more than 3:1), and as explained in Chapter 3, the wide spectrum of color signals means that color images can be compressed only slightly. As a result, medical image files tend to be rather large. However, advanced magnetic storage systems and high-speed LAN and B-ISDN can handle such files adequately. Therefore, we can expect that electronic media will spread rapidly in the field of medicine.

8.3 MUSEUMS

All digital SHD imaging systems have the ability to display images of quality and have the capacity to link various items organically to the image files. This capability will result in a new electronic catalogue systems for art and other types of museums. This does not mean that the image files will become substitutes for the real thing. However, they can be used for research, as is done on copies of expensive ancient artwork, especially when the artifact is fragile. In this section, we look at the significance of the electronic catalogue and the importance of SHD imaging in this exciting new field.

8.3.1 The Significance of the Electronic Catalogue

8.3.1.1 Ease of Accessing Information

The great promise of the electronic catalogue is the ease with which it permits information to be accessed. Human beings have long recognized the value of being able to access information quickly. This can be done with text, but not all human endeavor can be reduced to text, such as art, music, and the vast range of artwork assembled by archeologists and other collectors in museums.

The fact remains that text offers by far the easiest and most democratic means of making information available. If a book is well-organized, indexed and designed for convenient storage, the information can be accessed readily. The switch from scrolls to books centuries ago was motivated partly by the need to access information quickly, and partly because books take up less storage space. The significance of this advance becomes plain when we consider the hopelessness of using a dictionary if it were published as a scroll. The random-access approach to knowledge offered in a bound dictionary can be simulated in other books by the addition of an index, but since indexes can be rather arbitrary, this solution is unsatisfactory.

Certain types of information, such as images, are intrinsically best delivered in random-access form. There is often no logical way to order a set of still pictures, yet some sort of order must be assigned when images are printed whether as a scroll or a book. The arbitrariness of most ordering criteria becomes evident if we examine almost any museum catalogue, collection of photographs, or illustrated reference book. The usual type of ordering criteria—for example, by name alphabetically, by date of creation, by author, by geographic region—derives from our need to access the data, rather than from the meaning of the picture itself. The same can be said of textual information. Sorting literary compositions by author is reasonable, but the organization of historical archives and other documents is completely arbitrary except that they are based on time sequence.

8.3.1.2 Digitization of Image Media

Digitized media are vastly more amenable to random access than books, and thus potentially more useful. This usefulness is still not widely appreciated due to the deeply ingrained habit of using books. Image data, while enjoying the advantage of easy availability through random access, have not been digitized as quickly as text data for the following reasons.

1. Image files are much larger than text files.
2. Correcting and editing image files requires artistic sensibilities that are not necessary for similar operations on text files.
3. It is often technically difficult to manipulate high-quality image data.

4. In practical terms, it is impossible to abstract the "meaning" of an image file for cataloguing or indexing.
5. Several technical problems remain to be solved before SHD color displays become widely available.

Each reason may be explained briefly as follows. (1) The large size of image files increases the load on hardware and thus raises costs. (2) Since correcting and editing image files is a subjective task, it is difficult for managers to give clear instructions to subordinates, as can be done when text is edited. Even if an explicit instruction such as "Make the water droplets look as though they are about to fall," or "Tone down the color," is given, it will be executed differently by different operators. The skill of the editor depends heavily on his or her aesthetic sensibilities. (3) The technical difficulties will be readily understood by anyone who has ever dabbled in photography. These difficulties mainly concern the problems of lighting and focus. Use of correct lighting requires considerable experience. (4) Images cannot be retrieved by searching for a string of characters. Captions may be provided for this purpose, but they are subjective and thus do not support the image search well. (5) Many factors must be considered to provide the correct environment for accurate color reproducibility. For example, the type, placement, and intensity of light sources in the room where the display device is kept must be chosen carefully. The phosphors of the CRT or the type of backlighting used for the LCD display, the characteristics of the D/A converter used with the frame memory, and even connection cable characteristics, all make a difference.

As integrated circuits become more powerful, all of these problems are being solved, with the exception of item 4, abstracting the "meaning" of the image file. Obviously, this problem is neither hindered nor helped by hardware developments. In fact, it is not a serious impediment to the use of digitized images and its solution can be deferred.

The advantages of digital images are given below. These characteristics are extremely valuable to art and other museums.

1. Digital images are easy to reuse.
2. Digital images do not alter over time.
3. Digital images are readily duplicated and distributed.
4. Digital images are easily accessed.
5. Digital images offer freedom of layout.
6. Digital images offer flexible display for a variety of applications. Display format is not fixed.
7. Digital images may be accompanied by audio information.
8. Multilingual presentations can be easily prepared.
9. GUIs can be used for more "lively" explanations.
10. Video and still images can be linked.
11. Digital images are relatively easy to store.

These advantages are so obvious that no explanation is necessary. They greatly facilitate the implementation of new types of electronic museum catalogues.

8.3.2 Electronic Catalogues and SHD Imaging

In practical terms, an electronic catalogue can be thought of as a combination of a guidebook, a list of holdings, a dictionary, an art book, and a laser disc/CD-ROM (see Figure 8.3). Table 8.1 compares the advantages of various types of existing catalogues. A rating of "excellent" appears only once in each column clearly showing that each existing medium is able to accomplish only a single purpose. Obviously, this problem arises from technical issues and not from the intention of the producers of each type of medium.

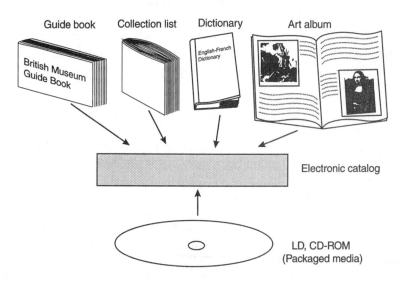

Figure 8.3 Concept of electronic catalogue.

Table 8.1
Comparison of Conventional Catalogues and the Related Media

	Guidebook	List of Holdings	Dictionary	Art Book	LD/CD-ROM
Quick grasp of essential data	Excellent	Fair	Poor	Fair	Good
Inclusiveness	Fair	Excellent	Excellent	Fair	Fair

	Guidebook	List of Holdings	Dictionary	Art Book	LD/CD-ROM
Image quality	Fair	Fair	Poor	Excellent	Fair
Ease of reuse	Fair	Poor	Poor	Fair	Fair
Ease of Access	Good	Poor	Poor	Poor	Excellent

Many persons interested in the arts would love to have a complete list of all the holdings of all major art museums around the world, accompanied by quality reproductions of each artifact and easy access to any record in the list. Such a dream catalogue is impossible using conventional media. There is not even a complete photographic record of all items in the Louvre museum. The enormous expense, the huge collection of photographs, and the high cost of reproduction would limit the number of subscribers. Naturally, every museum prints a partial catalogue of its holdings, but the question of which items should be included and which excluded is always a problem that permits no easy solution. If a medium existed that rated "excellent" for every characteristic in Table 8.1, its value would be immense, and the range of users would extend beyond the education and publishing establishments to which they have hitherto been confined and reach out to an unimaginably vast new audience.

The ideal electronic catalogue imagined above does not yet exist. It can exist only on electronic media, but the quality of existing digital images is still too poor to implement such an ambitious scheme. Yet this promise may be fulfilled by a fusion of SHD images with B-ISDN and computer technology. This is what many people are already referring to as the *electronic museum*. Obviously, problems of intellectual property, including copyright issues, must be settled before this can become a social as well as a technological possibility, but these problems can be solved by current encryption technology, including digital signatures, in conjunction with laws that reflect an understanding of the technological issues.

The electronic catalogue imagined at the beginning of this section is an extension of current catalogues and lists of museum holdings, and is intended to eliminate the inconvenience and inadequacies of these traditional publications rather than to substitute for the art objects themselves. The usefulness of such data can be appreciated by anyone who has combed the immense collections of the British Museum in London, the Louvre in Paris, the Smithsonian in Washington, D.C., and the Deutsches Museum in Munich, only to get tired feet and run out of time. Such persons can appreciate the usefulness of a tool that answers the questions of what to see and where to see it.

The European Vasari project arose as a means of creating an accurate, durable record of art objects using digitized image data, since such records do not deteriorate over time. This was completed in 1992 and was succeeded by the Marc project [2,3]. These projects used around 5,000 × 5,000 pixel images, a resolution that exceeds that of SHD images. However, the frame memory developed by the authors can scroll and zoom four frames of SHD images. The aesthetic value of this was recognized by numerous luminaries from the academic and artistic worlds during a session in which frames of 2,000 × 2,000 pixels were shown on a 28-inch CRT (screen size approximately 50 × 50 cm).

8.4 PRINTING

8.4.1 Digital Images in Conventional Printing

Completely digital color still image files exceeding 4 million pixels are already in use by the printing industry; this field can thus be said to be using SHD image technology. However, current use of digital images is limited to the intermediate stage of creating the process film, rather than an integrated system in which digitized image data are delivered from art houses and clients. Thus, the use of digital images by the printing industry has developed its own peculiarities.

Printers use a drum scanner to obtain RGB data, which are then converted to CMYK data, a process that is completely digital. The processes for converting from RGB to CMYK data are based on sophisticated knowledge of how the CMYK data will ultimately be expressed by the actual inks available to printers. Inks in the United States are being standardized to the point where the digital image data set can be relied on to produce an identical printed page, regardless of the inks used. Japanese industry officials are studying the issue. Standardization of inks will result in the widespread use of digital image data and will allow design studios to print proofs on their laser printers, which will be essentially the same as the final output of the printing press, offering significant efficiencies for both the design and printing industries and hastening the trend towards totally digital throughput.

Printing technology has been developed with a view to creating attractively printed pages, sometimes at the expense of the faithfulness of color reproduction. This fact is reflected in the characteristics of the drum scanner, which was designed in response to these demands. The composition of paper, the usual printing medium, the chemistry of the inks, and the mechanical design of the printing presses are all subject to more variability than electronic devices, which results in unavoidable discrepancies.

Printers have conventionally measured resolution in terms of the lines per inch of the screen, resulting in a different conceptual approach from that of dots per inch used with computer printers or CRTs. It is hard to fix the relationship precisely, but they are certainly related. For the purpose of comparison, 85 lpi is usually used for newspapers, 150 lpi for books, and 300 lpi for art books, with the caveat that great variation exists. Whereas ordinary laser printers have a resolution of 300 to 600 dpi, those used in the printing industry for making plates range from 1,200 to 1,500 dpi. Roughly speaking, 160 dpi corresponds to 10 lpi, but since the underlying concept is different, this correspondence should not be pushed too far.

8.4.2 The Use of SHD Images in the Printing Industry

The major printing companies of Japan have confirmed that an acceptable level of print quality can be obtained for the normal color printing of A4-size paper by using 2,000 × 2,000 pixel SHD image data as print data. By "normal" we mean offset printing rather than the gravure or intaglio printing typically used for art books. Given that SHD image data can be used for printing, how will this change the print industry? Clearly, it will revolutionize the printing process by giving more initiative to the designer.

In the present world of commercial printing, time constraints always force the designer to make certain compromises, because the images must be derived from photographs, which must be developed, fixed, dried, and digitized on a drum scanner. This limits the time available for creative work.

Clearly, the need to perform routine operations limits the amount of time available for creative work. The quality of the final artwork is almost always directly proportional to the time consumed in the trial and error of the creative process. Obviously, some exceptions exist, but generally the quality of the final artwork improves as more time is spent on its creation. The fact that digital processing creates more time for the designer will likely result in major changes in printing.

Obviously, the elimination of time-consuming tasks also makes it possible to design more artwork within the same time span. In other words, the introduction of SHD image technology has the potential to increase both the quality and the quantity of artwork available for printing. Since the technology for increasing quantity is already available, the contribution of SHD images will likely be made in the quality arena, and this will trigger many changes. Obviously, this includes vast changes in the organization of design departments.

8.5 DYEING AND PRINTING TEXTILES

The color reproducibility of SHD images is adequate for dyeing and printing textiles, so applications in this area are under consideration. An electronic changing room system, or *virtual dressing room*, which is an application of the electronic catalogue, is a likely application. The design itself would become more efficient, since the trial-and-error process can be done more quickly on line, as we already see in architecture and interior decorating. While the catalogue itself is no different from previously discussed electronic catalogues, there is one difference to the overall system in that the catalogue can represent a nonexistent inventory. The system thus depends completely on made-to-order products. One of the first widespread uses of computers was for stock quotations, since the product had to be made available to the purchaser at the earliest possible moment without the delay of printing a list. The situation in printed textiles is precisely the reverse. The products themselves are expensive and require some degree of tailoring or sizing, thus reducing their universality. It therefore makes more sense to advertise a "virtual inventory" and to manufacture the product only after orders have been received. The virtual dressing room simply refers to the system for placing orders from the electronic catalogue, similar to other uses of virtual reality (VR) for handling sales of made-to-order goods. Whereas VR, strictly speaking, heightens the sense of realism by simulating three-dimensional space, the SHD imaging system accomplishes this with high-resolution color reproduction. In the section below, we will discuss the application of SHD images for the virtual dressing room.

The Virtual Dressing Room

The question of whether three-dimensional imaging or high resolution and color reproduction is better for evoking realism depends on the product to be sold. Textiles can clearly benefit from high resolution and color reproduction, which have the advantage that they can be printed. However, like three-dimensional images, virtual dressing rooms can never be printed.

The following example is an application of the virtual dressing room to the sale of Japanese kimonos dyed in the traditional Yuzen style. First, the image of the customer wearing a white kimono is digitally captured. The plain white kimono, devoid of any pattern whatsoever, is inexpensive and can be stocked by all participating retailers. The customer then chooses a design. No actual dyed fabric of the design need exist, as long as it exists electronically.

The pattern data are mapped onto the white kimono data, giving an image more or less identical to a customer wearing an actual kimono of that design, as shown in Figure 8.4. This system is already being used in Japan. The composite image can be viewed from many different angles, although Figure 8.4 shows just one view. The quality of the image is such that only experts are able to distinguish the virtual dressing room from a photograph of the real thing.

Obviously, this system can be applied to any type of apparel. The kimono is the first item of apparel to be sold by virtual dressing room systems because it is extremely expensive and takes a long time to create. As such systems become less expensive, they will be used by haute couture designers, whose operation is also based on receiving special orders from clients. At some point, these systems will become inexpensive enough to be used throughout the fashion industry. In addition to reducing inventory and generating more orders, these systems make the designer's job more efficient.

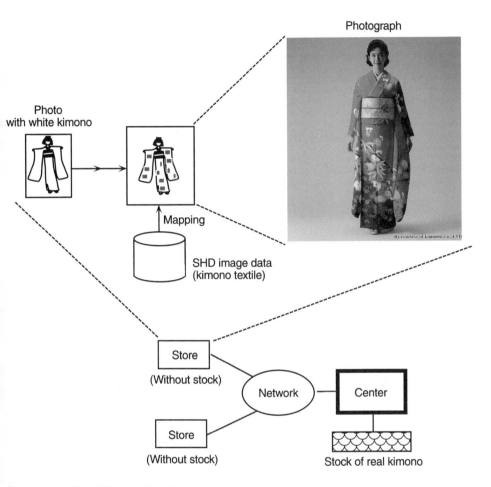

Figure 8.4 Virtual kimono dressing system.

8.6 PHOTO LIBRARIES

A library of photographs or films is an archive of motion pictures or video and still photographs that can be made available for viewing. It is no different from an ordinary library, except that reels of film and files of still photographs are stored instead of books. The resolution of SHD images is better than that of 35-mm film, so they can be used to archive the same data. SHD images are also useful for cataloguing and retrieving larger film formats, and the digitized data can be scrolled or zoomed efficiently. Thus, SHD images can be used to preserve photo libraries in semi-eternal form, and because digitized images are semi-eternal and can be readily duplicated in their entirety, restrictions on display or close viewing can be almost entirely eliminated.

8.6.1 Deterioration of Image Quality

Books last a long time, and although acids in the paper do cause problems with older books, the trend to using acid-free paper is increasing the durability of printed matter. The chemicals used in photographs are not as stable, so the deterioration of images is a constant problem for photo libraries. This problem is addressed by restricting the access to and display of photographs and carefully controlling the temperature, humidity, and air circulation of the storage facilities themselves.

Film base, conventionally made of celluloid but more recently of acetate or polyester, is the cause of much deterioration. The glass dry plates used by early photographers provide surprisingly clear photographs even after the passage of a century. In contrast, microfilm collections made thirty or forty years ago using standard commercial film base are already starting to deteriorate.

In contrast, SHD images are digital and so can be stored on magnetic disks, magneto-optical disks, magnetic tape, or other media. MO disks and magnetic tapes use a plastic substrate that is subject to the same problems of deterioration as film base, but if the data on these media are copied onto fresh media before they start to become corrupted (e.g., before the rate of read errors reaches a certain level), the digital data file can last forever.

Obviously, when data are to be copied at regular intervals, an error correcting code (ECC) should be used, allowing any physical errors occurring within the range of the ECC to be completely restored. This restoration capability is proportional to the amount of code added to the original data. The selection of an ECC depends on the way in which the medium (MO disk or magnetic tape) is likely to deteriorate. However, as the memory capacity of storage media continues to rise, practical limitations on the amount of code added to the original data become less severe, allowing decision makers to focus entirely on the ECC that best preserves the data. Occasionally errors occur that cannot be restored by the ECC; in this case, the close correlation between adjacent data in an image usually allows a near-authentic restoration.

8.6.2 Retrieving Image Data

Conventional film viewers for searching archives are extremely troublesome to operate and require the viewer to examine every image in sequence until the sought-after picture appears. This system is quite inefficient in terms of both time and effort. A digital system is incomparably easier to use, because there are no constraints on the order in which the film is viewed.

In summary, digital SHD image archives are superior to conventional film archives in their ability to preserve film (a fact that will gain increasing recognition by the public) and in the ease of retrieving, displaying, and viewing the images themselves. The availability of such digital archives through telecommunications systems will increase the number of patrons, thus making the archives more valuable and blurring the distinction between conventional libraries and film archives.

8.7 LIBRARIES

8.7.1 Conventional Libraries: Limitations and Prospects

Libraries are more than just collections of books, and their value to society cannot be overestimated. Libraries must be organized to allow patrons to identify and access the desired books efficiently. This purpose is accomplished by means of a cataloguing system and by the practice of keeping the books in the locations specified by the catalogue. In conventional libraries, one may locate a publication in the catalogue only to find that it is on loan to another patron. In other cases, the desired book may not even be listed in the catalogue. Libraries can be thought of as databases, but in most cases only the card catalogues have been digitized, not the publications themselves, so they are very imperfect databases. This is because digitizing the contents of books presents technical problems that are extremely difficult to solve.

The most difficult technical problem has been the problem of storing the vast amount of data required for digital images of books at an acceptable level of quality. However, this problem is being solved as magnetic memory devices offer more capacity at lower prices, allowing entire books to be digitized. B-ISDN allows entire books to be transmitted in digital form to remote computers. Thus libraries are becoming more like other databases, which allow us to actually read the records once they have been retrieved.

What form should books take in such an electronic library? One way is to store the entire book as image data. A resolution of 300 to 600 pixels per inch (ppi) is adequate for display on a CRT and printing out hard copies. An entire newspaper page can be digitized by SHD imaging and still be quite legible. This is more than adequate to digitize a two-page spread of an open book.

Although digitizing the contents of books as SHD image data may require a lot of memory, it is a first step toward presenting library holdings in digital format. The memory requirement becomes less of an issue as storage devices continue to offer increased capacity at lower price. However, this approach is merely an electronic form of microfilm. Like microfilm, it does not allow patrons to search the contents by keywords, change the typeface or font size, or to perform other operations commonly used by the computer-literate public.

The alternative is to convert the text to character code (plus page layout information), representing only illustrations and graphs as image data. Not only does this reduce the size of the file, but it permits users to search by keywords and allows the typeface and font size to be changed. The book file thus resembles any other text file and can be used as such. The file can be used on many platforms if JPEG is used for graphics and photographs and PostScript is adopted as the page description language for the text, associated information, and page layout data. PostScript can even be used for some graphics [4,5].

The development of an electronic viewer for such PostScript-JPEG book files is an urgent desideratum. There are no technical problems obstructing the development of such a viewer. If file formats can be standardized, the software can be developed for workstations and PCs and be made available as public domain software. Obviously, superior versions of such software might be sold at a high price, but probably it will be no more expensive than telecommunications programs for PCs.

8.7.2 Electronic Libraries

The all-digital database of publications, together with the electronic viewer, may be thought of as an *electronic library*. The development of this concept will change society profoundly. Here we can only discuss two of the issues in depth: the transformation of the library itself and the problem of copyright protection.

The availability of electronic library holdings over B-ISDN will make the geographic location of the library irrelevant, allowing patrons to access books from the comfort of their homes. Just as people who use on-line banking services do not care where the bank's computer is or how many computer sites the bank maintains, so too governments and universities will have the freedom to put the physical library virtually anywhere; proximity to the rest of the campus will no longer be a consideration. The electronic library will also blur the distinction between library and bookstore, since it will be easy for patrons to download a book, duplicate the file, and print out hard copies. As the result is not much different from buying a book in a store, this will greatly change the role of bookstores in society.

Libraries are basically institutions that buy large numbers of books and make them available for free to their patrons. This process has not been perceived as threatening the copyright interests of authors, because library books cannot be readily duplicated. From an author's point of view, it would be nice to receive a royalty from everyone who uses his or her book. Obviously, the total royalties received by an author are diminished by the fact that libraries make single copies of books available to multiple users. This factor is offset by the fact that library purchases guarantee that a certain number of books will be sold. The balance thus achieved will be upset by the ease with which electronic books can be copied. The combination of copy machines and libraries already acts to reduce book sales. In some countries, the price of a copy machine already includes a fee distributed to authors to compensate them for lost royalties. Obviously, unless the titles of the most frequently copied books can be ascertained, which is impossible to do accurately, the distribution of such royalties is inherently unfair.

We should remind ourselves that, in addition to copyright, there are other issues at stake, such as protection of our natural resources by reducing the use of paper and the preservation of books and other publications as valuable cultural properties. However, as the electronic library begins to displace the function of the bookstore, society must develop a new system to protect the rights of authors.

The use of digital signatures offers a technical solution to the problem of copyright, but it does not address all possible cases. A determined, technically competent individual can still infringe with impunity. Digital signature technology merely prevents simple cases of copying that do not require technical know-how, such as is done with CDs. It is not clear that copyright can be completely protected by designing hardware systems to prevent infringement. Even if it could be done, the cost and inconvenience of such technology may well be prohibitive. The authors feel that, rather than enforcing purely technical solutions, it is easier to increase the awareness of copyright in society at large and convince the public that infringement is a serious crime, as well as making it more difficult to perpetrate such crimes.

The technology already exists for creating a system whereby royalties can be collected from individuals who duplicate copyrightable materials. For instance, by assigning a unique ID number to every computer, and registering this number every time a copyrightable file is downloaded to that computer, royalties could be deducted from the owner's credit account. The cost of such a system is much less than that required to completely protect authors from infringement, but obviously it cannot prevent the owner of the computer from making duplicates once the file has been downloaded. However, this problem can only be addressed by raising social consciousness. This problem is not limited to books, but includes video tapes, CDs, photographs, illustrations, computer programs, and any creation that has been or can be digitized.

If this problem can be solved, electronic libraries will soon play a larger cultural role than their conventional counterparts, embracing the functions of art and natural history museums, concert halls, archives of all sorts, and film libraries. This is the promise of multimedia. However, there is also a dark side in that it becomes easier to manipulate information, requiring more diligence on the part of the user. Unless we remain a society dedicated to the faithful recording of facts as facts, we may descend into a morass not unlike the dystopia presented by George Orwell in *1984*.

SHD images will play a major role in the electronic libraries of the future. Currently, CRTs are large, heavy, and anything but flat, and certainly not the ideal tool for reading a book. However, the LCD displays that will become common within the next decade will make this possible. In an aging society, the ability to change font size will allow many people to continue reading books in spite of declining visual acuity.

Earlier it was stated that the electronic library will blur the distinction between libraries and bookstores; but as the electronic library expands to embrace the functions of art and natural history museums, concert halls, archives of all sorts, and film libraries, it will also change the role of video stores, record shops, software and video game vendors, and other merchants. Newspapers will also be affected. Although the shape of the future electronic newspaper is still uncertain, we can reliably assume that the organization itself will resemble a communications company. Television stations and other mass media will also change.

8.8 MAPS

Maps are probably one of the most detailed forms of graphics in common use, and their utility is widely appreciated. However, detail in and of itself is not always valuable and sometimes prevents us from understanding the spatial relationship of large-scale features on a map. An ideal map would allow us to select the degree of resolution (scale) and type of features (symbols) to be displayed, allow us to frame only the area relevant to our curiosity, and perhaps even place the feature most important to us at the center. Obviously, this ideal can never be realized on a printed map, which is one reason why there are so many different kinds of maps. Yet there is a limit to the number of different types of maps that can be made.

Digital maps can satisfy the requirements of the ideal map and are thus rapidly gaining in popularity. However, conventional display systems lack the resolution necessary for maps and thus compare poorly to the visual appeal of conventional printed maps. Anyone who has seen a global positioning system (GPS) system in a car will probably agree. The coarseness of these maps is tolerated because of the usefulness of the accurate guidance system, but this is an exceptional case that only applies to satellite-based automotive navigation.

Map files produced by SHD imaging systems show none of the coarseness of conventional digital maps and are as pleasing to the eye as their printed cousins. In fact, such maps are already being used to good effect for research and education. SHD imagery also allows windows to be opened against the background of the map and photographs and other close-up information to be placed inside, thus significantly increasing the utility to researchers and educators. Obviously, video and audio may be added to these windows, creating a full-fledged multimedia system.

8.9 EDUCATION

The virtual classroom is likely to be the major application of SHD imaging in education. The ideal student-to-teacher ratio of 1:1 is economically impossible,

Figure 8.5 A layout of a classroom designed for use of SHD screens.

and even if it were possible, not enough trained educators are available. These conditions are reflected in the way modern school systems have developed. Even so, there are not always enough skilled teachers. If the purpose of education is to disseminate knowledge and critical thinking, we can safely say that SHD imaging will greatly assist in the dissemination of knowledge, since the quality of the images appeals to both the emotional and the rational components of the human character. Obviously, it also fosters critical thinking to a certain extent, but since this faculty can only be fully developed by responding to a variety of interpersonal situations, it is best addressed as a dialogue between individuals.

Before the development of printing, most higher education took the form of debate. With the advent of moveable type, inexpensive textbooks became available and the dissemination of knowledge has come to predominate in education to the extent that many claim that the teaching of critical thinking has been short-changed. It may be that the development of interactive education will serve to correct this trend.

Although we cannot predict all the ways in which SHD imaging will affect education, it is already seen as the answer to many long-standing problems. Figure 8.5 shows the layout of a classroom built around a central screen for SHD imaging. Obviously, this screen can also accommodate images from workstations and PCs.

The only thing that can be said for certain is that although the role of the teacher will remain unchanged, he or she will no longer use the same tools. The teacher will not need to prepare in advance video representations of material that will be written on the blackboard, nor will characters on the blackboard be required to be oversize in order to be legible to the video audience. Futile attempts to televise blackboards on which the teacher writes with artificially large letters will be happily abandoned. The educational promise of television, never fully realized, has at least generated many sound theoretical approaches that can now be implemented by SHD imaging and other technologies.

8.10 APPLICATIONS FOR SHD MOVING IMAGES

In this chapter so far, we have mainly discussed applications for SHD still images while barely touching on the subject of SHD moving images. This is because there are few technical difficulties in the application of SHD to still images and therefore, in the near future, we expect to see this area come into its own. In contrast, the use of SHD moving images requires further advances in TV cameras and CODECs, leaving this field as yet in an experimental stage. Another unresolved issue regarding the use of SHD moving images is the lack of any popular existing moving-image media of comparable quality except 70-mm movies. SHD still images, on the other hand, are supported by existing media such as printing and film, which are used in all forms of intellectual life in our society. The application of SHD still images, therefore, is based on the use of existing media.

While there are existing forms of moving-image media, including movies and TV, their main purpose is for entertainment and news. Recently, they have also been used for educational purposes, but due to problems with the media's image quality, this has not spread very far. The goal of HDTV is to improve image quality, but the level of quality it attains is only high enough for entertainment applications; therefore, it will not have enough of an impact to be used for other purposes. This, however, does not mean that SHD moving images will form a precedent. "Beyond HDTV," which is part of the title of this book, means that we expect SHD imaging to reach beyond existing moving-image media, primarily entertainment and news, but it is difficult to discuss with much clarity exactly when and how this will be achieved. In this section, we will discuss applications for moving imaging, even if it deals with such uncertain issues.

8.10.1   Summary of Technological Advances

As discussed in Chapter 3, the standard MPEG2 algorithm can compress an SHD moving-image sequence that would normally require 6-Gbps transmission rate by 1/40, allowing it to be transmitted at 150 Mbps. This level of compression yields an SNR value of 40 dB, which means that the loss in image quality can scarcely be detected. This is good news for storage devices and also means that existing digital data recorders that operate at a maximum of 256 Mbps can be used for recording moving images. Of course, sending and storing SHD moving images requires CODECs that can perform MPEG2 algorithms in real time. As described in Chapter 6, advances in DSP system architecture and VLSI technology have raised the prospect of this becoming possible, although the equipment required is currently rather massive. For example, by using an array of recently announced parallel processing DSP chips such as the TMSC80, we could design a CODEC that operates in real time. (Its programming is quite difficult.) The advances in CCD image recording elements for SHD video cameras are remarkable, and while costs are still rather high, the technology will probably reach the developmental stage within three years. Under such conditions, it seems that the requisite technology to handle SHD moving images in various applications is advancing steadily.

8.10.2   Movie and Entertainment Video

SHD imaging can most naturally be applied to the field of making movies. SHD moving images rival the quality of current 70-mm motion pictures without using film. Eliminating the need for film will dramatically lower the cost of making 70-mm motion pictures. It costs approximately $100 to develop 1 second's worth of 70-mm film, while the cost for storing the equivalent amount of SHD moving images is estimated to be only $10. This is reasonable if you imagine the price of a digital audio tape or an 8-mm video tape, which can store 1 GB of digital data, enough for storing 1 second of SHD moving images. In addition, editing recorded SHD images is much easier than film editing and lower in cost; the medium can be reused if necessary.

SHD moving imagery will even change the way images are filmed and edited. For example, current opera laser disks cannot adequately reproduce the true essence of the performance. This is due to the fact that the present TV system, on which laser disks are played, cannot display the singers' performance at the same time as showing the stage and props that cost a great deal of money. Both factors are vital in appreciating the opera. SHD will likely result in changes to the way in which movies and documentaries are made and, as we will discuss later, will even have an impact on the way educational image media is edited.

8.10.3 Surveillance and Observation Systems

The fields of surveillance and observation demand systems that can produce images with high spatial resolution. The end result of these systems must be a still image of extremely high spatial resolution. Unfortunately, because it is not known ahead of time exactly when the desired image can be recorded or what its position will be, such systems must record continuously or, more precisely, capture and store moving images. Since, however, they do not have to provide a high frame rate, slow scanning (low frame rate) can be used. Surveillance systems in the future will likely use SHD imaging with low time resolution.

8.10.4 Medical Imaging

As we stated previously, the application of SHD still imaging to the medical field looks quite promising. The relationship, however, between moving images and medicine is not so simple, because there are so many diverse aspects therein, including medical education. If, for example, technical difficulties regarding SHD cameras can be resolved, SHD imaging would probably be widely used to record operations for the purpose of training surgeons. Another aspect is telemedicine including remote diagnosis. X-rays, MRIs, CTs, and other still imaging methods are already used in remote diagnosis. To faithfully reproduce the situation of a doctor physically examining a patient, remote diagnosis systems that use SHD imaging could be used. A number of tests have been conducted using normal TV and HDTV for recording surgeries or remote diagnosis, but the results have been poor due to the conventional media's low spatial resolution and inability to faithfully reproduce colors. It is therefore conceivable that SHD moving imagery could be applied to the medical field. How SHD moving imagery will be used is likely to be influenced by what level of prominence it gains within medical systems.

8.10.5 Electronic Museums

The use of SHD imaging for electronic museums will be primarily for still images. However, some subjects with guide systems that provide narration will require moving images to fill in the gap that occurs when images are suddenly changed, allowing fade-ins, fade-outs, and wipe-outs. These kinds of operations can only be achieved using moving images. It does not require, however, that all sequences be saved as moving images. The moving-image portions could be generated using digital processing, which could be inserted between the various cuts.

8.11 HYPERMEDIA

Hypermedia and multimedia are much overused buzzwords, about which we can say little for certain except that they are best used to describe to future applications of technology rather than presently available systems. Even so, the underlying technology for implementing these visionary projects is already in place or soon to become available. The discussion given below focuses on these close-at-hand technologies and how they must be improved in order to give substance to the claims of visionaries that profound changes are about to sweep over society.

8.11.1 Characterization of Hypermedia

In order to clear the air of buzzwords, let us ask the question: What characteristics do we want to see in a fully realized hypermedia system?

1. It should be capable of manipulating high-definition image data, including video.
2. It should provide integrated access to databases.
3. It should be interactive.
4. It should permit two-way communications.
5. It should provide full functionality for recording as well as viewing.
6. It should run in real time.

Each of these requirements aims at instant access to useful information, something that conventional media falls short of doing perfectly. Multimedia and hypermedia are put forward as candidates for accomplishing this objective.

Each type of existing medium fails in a different way to fulfill the promise of hypermedia, as shown in Table 8.2. Note that ratings other than "excellent" are not shown. Note also that the table only examines the properties of each medium by itself. There are many cases where the medium can be significantly enhanced by another device, but this was disregarded, since it would hopelessly complicate scoring. For example, the television receives no marks in the column "full recording functionality." Obviously, a recording can be made with a VCR, but unlike other media such as faxes and photographs, the recording function is not an intrinsic part of television as a medium.

PC communications rate "excellent" in most categories, and some networked PCs could also rate "excellent" in real-time performance (for example, those linked to the same LAN network). The only thing that PCs lack is high-quality video. Networked PCs that are capable of real-time processing and that can manipulate high-definition images, including video, will satisfy all of the qualifications for hypermedia. Therefore, of each type of media shown in Table 8.2, networked PCs are the ideal candidate for hypermedia.

Table 8.2
Adequacy of Existing Media

	High-Quality Video	Access to Databases	Interactive	Two-way Communications	Full Recording Functionality	Real-Time Performance
Printed materials					Excellent	
Photographs					Excellent	
Movie film	Excellent			Excellent		
Telephone				Excellent		Excellent
Radio Broadcasts						Excellent
Television						Excellent
Compact discs					Excellent	
Fax machines					Excellent	Excellent
Networked personal computers		Excellent	Excellent	Excellent	Excellent	

Does this mean that the ideal hypermedia platform can be realized by giving networked PCs the ability to manipulate high-resolution images? The authors answer a resounding "Yes!" to this question. However, this endorsement is based on the assumption that the current networked PCs will change form drastically and manipulate high-resolution image data. This can only be implemented by the development of new protocols for handling high-speed video communications, along with new routers, the central element of PC communications ports and nodes, to handle these protocols. However, there is another problem even more important that these technical issues, as is discussed below.

8.11.2 The Most Basic Stumbling Block for Hypermedia

Hypermedia (or multimedia) will probably be developed for the purpose of accessing required data quickly, where "required data" include any form of data, even video, as well as audio and text data. Furthermore, we predict that in future hypermedia systems, the text data will not be a single monospaced font as is currently used in PC communications, but will be a fully formatted, aesthetically pleasing document no different from a book or magazine article. There is some doubt about whether hypermedia can completely satisfy the requirement of making all types of data—including video—instantly available. We discuss this issue in some depth below, beginning with the history of printing itself.

The invention of moveable type made textual data widely available at low cost. The early bookmakers did not understand the implications of this and instead printed imitations of the illustrated manuscripts of the Middle Ages. These manuscripts were works of art and could not hope to become "widely available at low cost." Then came the Protestant Reformation. The early reformers seized the opportunity to make their arguments "widely available at low cost," and in doing so they transformed the nature of printing.

Printing became a means of disseminating information quickly; in other words, of making required data available as soon as possible in as many different places as possible. Although we are in the habit of thinking of modern libraries as heirs to the ancient library at Alexandria, our modern libraries lack the royal patronage that fostered their ancient counterparts, and are therefore dependent on books that are "widely available at low cost." Once printed books became available, the number and size of libraries began to increase dramatically, making vast quantities of information instantly available to many people. This had a tremendous impact, whether we are talking about the vast collection of a university library, a small town library, or a shelf of books in a home.

For all its contributions, there are limitations to print technology. Once a manuscript has been bound and published, the contents cannot be updated or corrected without publishing a new edition. Books must be transported from place to place, take up space when they are stored, and require constant effort to keep organized.

Now that advances in telecommunications allow us to transmit data over vast computer networks, we no longer need to collect books and journals in order to have information at our fingertips. Required data can be requested from distant locations, and documents can be updated and corrected with a minimum of effort. Obviously, there are many different kinds of databases, the most pervasive being on-line banking services and reservation systems for airlines and trains. These areas are logical candidates because instant availability of up-to-date information is required in these industries.

However, many problems arise when we try to design an on-line system, such as is used for banking and reservations, that also permits us to handle image data. Simply put, databases that accommodate image data must provide the following.

1. The data provided cannot be limited to text.
2. The request for data may not be expressed in a clear, logical format specifying the sought-after data.

Conversely, the limitations of current database systems are:

3. Current databases only handle text data.
4. Current databases require that the query for required information be specified in a clear, logical formula.

Not all data can be converted into text format. Although existing databases are still extremely useful to society, we should not over-rate them, for most human intellectual activity, whether rational or emotional, can only be satisfied by conditions 1 and 2.

Obviously, future systems that satisfy conditions 1 and 2 will necessitate new hardware and require greater memory capacity at lower cost to handle the size of image files, which are an order of magnitude larger than text files, but this problem is already being solved.

The most difficult problem is that of image retrieval, since it is almost impossible to determine the meaning of a picture. This problem is sometimes encountered in text data, but to a lesser extent. How can we write a logical formula to retrieve a specific picture? Obviously, we can write a caption for each picture and then search for certain keywords. However, the captions would be subjective determinations of the meaning of each picture, and would require a great deal of work to write. Furthermore, this would be impossible to implement for every frame of a video.

Attempts to identify certain compositional features and link them to databases in a hypertext (note that this is different from the use of icons and other "hot buttons") have all failed, indicating the futility of this approach. Every attempt to assemble a database of images is confronted by this problem. Objective data such as the date a photograph was taken, the photographer's name, location where the photograph was taken, the type of film used, can be assembled, but going beyond this strictly limited approach is fraught with problems. It is completely impossible to describe the entire meaning of a picture, so a part must stand for the whole, and the selection of a representative part will always be subjective.

We can only conclude that any hypermedia system that makes any type of requested data, including video, available immediately will be fundamentally different from text-based systems. Although we cannot predict what form hypermedia will take in the future, the authors anticipate that it will be based on high-resolution browsers that will permit relatively efficient selection of graphics files from distant databases. Obviously, SHD imaging technology is one solution for the provision of high-resolution image files, and B-ISDN offers a method of transmitting large image files over long distances.

8.12 SUMMARY

In this chapter we have discussed the future of SHD imaging. SHD imaging spans a number of fields, and the technical level of the applications considered is still uncertain in many cases, all of which makes it difficult to offer precise predictions. Reasonable projections may be made for areas of strong social concern, such as medical imaging and education, or areas such as topography, which are entirely dependent on graphics processing. However, the authors feel that the unknown potential of this new technology holds great promise, so we have deliberately avoided imposing a false sense of order on future applications of this technology. The difficulty of predicting such future applications implies that the combination of SHD imaging with other media will have a profound impact in ways that we cannot foresee.

References

[1]   Bushberg, J.T., J.A. Seibert, E.M. Leidholdt, Jr., and J.M. Boone, *The Essential Physics of Medical Imaging*, Baltimore: Williams & Wilkins, 1994.

[2]   Martinez, K., "High Resolution Imaging of Paintings: the VASARI Project," *Microcomputers for Information Management*, Vol. 8, No. 4, Norwood, NJ: Ablex Publishing Corp., Dec. 1991.

[3]   Derrien, H., "The MARC Project: A New Methodology for Art Reproduction in Colour," *Proc. Electronic Imaging & The Visual Arts '94* (EVA'94), Paris, April 1994.

[4]   Adobe Systems Inc., *PostScript Language Reference Manual*, 1985.

[5]   Pennebaker, W.B., and J.L. Mitchell, *JPEG Still Image Data Compression Standard*, New York: Van Nostrand Reinhold, 1993.

Appendix A:
Actual Resolution Examples

Figure A.1 shows the actual resolution measured for different types of film and lens systems. This figure shows the relationship between peak-to-peak contrast and equivalent resolution in TV lines calculated from the developed test chart. The test chart used is shown in Figure A.2. The measurement was done using the counterpart of the chart. A Fujichrome Velvia (ISO50) was used as the film for all sizes. For 35-mm film, the lens system Nikon F4 (35- to 70-mm) Zoom Nikkor was used. A Hasselblad Distagon 50-mm was used for 6 × 6 film. A Zenza Bronica GS-1 was used for 6 × 7-mm film. Professional users must be prepared for significant disparity in the actual resolution when the performance of the camera is considered in conjunction with the resolution of the film itself. In this figure, resolution measured using a digital still camera (see Chapter 4, Section 4.5) is also shown. The Hasselblad Distagon 50-mm was used for this measurement. The current digital camera provides similar resolution with 35-mm films if a large lens system is used.

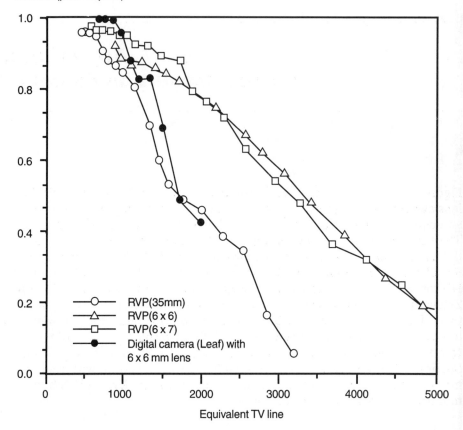

Contrast (peak-to-peak)

Equivalent TV line

Figure A.1 Actual resolution measured for various film and lens systems.

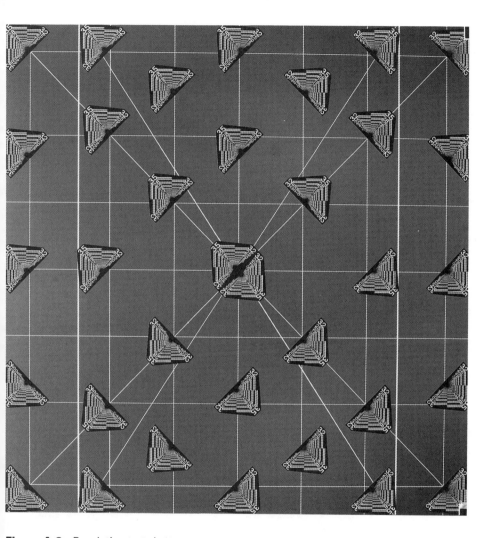

Figure A.2 Resolution test chart.

Appendix B:
The WYSIWYG Color-Matching Technique

As discussed in Chapter 7, the authors have used this technique in our experimental SHD image system to match the color of digitized images displayed on a CRT with the color of the original photograph, recorded on film. The technique is outlined as follows.

1. Display the original photograph, u, with a slide projector at the same size as the image displayed on the CRT. The image should contain a color patch to serve as a basis for color comparison.
2. Measure the optical spectrum data (S_i (l), $S_i'(l)$) for both images.
3. Apply the CIE color-matching function to the measured S_i and S_i', and obtain the tristimulus values for each color patch.
4. Convert the derived tristimulus values into signal values c_i, c_i' in the RGB or other color space.
5. Obtain the matrix $\mathbf{M}$, which converts c_i' in such a way as to minimize, under an appropriate metric, the color difference in color space. The least squares might be used as a metric; in this case, the color difference E in color space is

$$E = \sum_i \| c_i - Mc_i' \|^2 \rightarrow \delta \qquad \text{(a-1)}$$

6. Use $\mathbf{M}$ to produce a color-corrected digital image. The color-corrected digital image $\mathbf{v}(x,y)$ is derived from the color transform function g and $\mathbf{M}$ as follows:

$$v(x,y) = g^{-1}(M g(u'(x,y))) \qquad \text{(a-2)}$$

Glossary

A

A/D	analog to digital
ALU	arithmetic and logic unit
ASIC	application-specific integrated circuit
ATM	asynchronous transfer mode
ATV	advanced TV

B

B-ISDN	broad-band integrated services digital network
BiCMOS	bipolar complementary metal oxide semiconductor

C

CCD	charge-coupled device
CCITT	Consultative Committee in International Telegraphy and Telephony
CD	compact disc
CIE	Commission Internationale de l'Eclairage (International Commission on Illumination)
CMOS	complementary metal oxide semiconductor
CMYK	cyan, magenta, yellow, and black
CPU	central processing unit
CRT	cathode ray tube

D

D/A	digital to analog
DCT	discrete cosine transform
DFT	discrete Fourier transform
DMA	direct memory access
dpi	dots per inch
DRAM	dynamic RAM
DSP	digital signal processor
DVI	digital video interactive

E

ECC	error-correcting code
ECL	emitter-coupled logic
EMF	electromagnetic field

F

FCC	Federal Communications Commission
FD	floppy disk
FDDI	fiber distributed data interface
FPGA	field-programmable gate arrays

G

GP	graphics processor
GPS	global positioning system
GUI	graphical user interface

H

HDTV	high-definition TV
HiPIPE	highly parallel image-processing engine

I

I/O	input/output
ICE	in-circuit emulator

IEICEJ	Institute of Electronics, Information, and Communication Engineers of Japan
ISDN	integrated services digital network
ISO	International Standards Organization

J

JPEG	Joint Photographic Experts Group

K

K-L	Karhunen-Loeve
kTr	kilotransistor
kVA	kilovoltampere

L

LAN	local-area network
LCD	liquid crystal display
lpi	lines per inch
LSI	large-scale integration
LUT	lookup table

M

MC	motion compensation
MO	magneto-optical (disk)
MPEG	Motion Picture Experts Group

N

NTSC	National Television System Committee
NTT	Nippon Telephone and Telegraph

O

OS	operating system

P

PAL	phase alternative line
PC	personal computer
PE	processing elements
PGA	pin grid array
ppi	pixels per inch

Q

QMF	quadrature mirror filter

R

RGB	red, green, and blue

S

SCID	standard color image data
SCSI	small computer system interface
SECAM	sequential color and memory
SHD	super high definition
SNR	signal-to-noise ratio
SRAM	static random access memory
STFT	short-time Fourier transform

U

ULSI	ultralarge-scale integration

V

VA	voltampere
VBS	video burst, synchronous
VLC	variable-length coding
VLSI	very-large-scale integration
VOD	video on demand
VP	vector processor
VQ	vector quantization

VR   virtual reality
VSP   video signal processors

Y

YIQ   color signal representation recommended by the International Telecommunication Union
YUV   color signal representation of the National Television System Committee (NTSC)

About the Authors

Sadayasu Ono

Sadayasu Ono received his B.S., M.S., and Ph.D. degrees in electrical engineering from Keio University, Yokohama, in 1971, 1973, and 1976, respectively. In 1976, he joined the Yokosuka Electrical Communication Laboratories of Nippon Telephone and Telegraph Public Corporation (NTT). He was a member of the design team that developed the NTT digital filter CAD system and the first floating-point DSP LSI, and he became a project leader of the first C compiler development group for the DSP. During the last several years, he has been responsible for research and development on the ultrahigh-speed parallel signal processor for SHD image signals. Currently, he is a group leader of the Super Signal Processing Research Group at NTT Optical Network Systems Laboratories.

He is a coauthor of the two books *Software on Signal Processors* published by Corona Co. in 1989 and Digital Signal Processors published by Ohm Co. in 1990.

Dr. Ono is a member of the IEEE, IEICE, IPSJ, and is associate editor of *IEICE Transactions*. He is also the vice chairman of the Digital Signal Processing Study Group of the IEICE.

Naohisa Ohta

Naohisa Ohta received his B.S., M.S., and Ph.D. degrees from Tohoku University, Sendai, Japan, in 1973, 1975, and 1978, respectively. In 1978, he began work at NTT's Yokosuka Electrical Communication Laboratories, specializing in digital speech signal processing for communications. Since then he has been engaged in research and development of speech/audio signal processing, image/video signal processing, parallel DSP systems, packet video transmission technologies through ATM networks, super-high-resolution digital images, and VLSI implementation for programmable communication systems.

He is currently a senior research engineer and leading a research group at NTT Optical Network Systems Laboratories. His research interests include parallel signal processing, SHD image processing, high-speed VLSI architecture for digital communications, and high-level design methodologies for communication systems.

He is a coauthor of three books on digital signal processing and an author of the book *Packet Video: Modeling and Signal Processing.*

Dr. Ohta is a senior member of the IEEE, a fellow of the SPIE, and a member of the Institute of Electronics, Information, and Communication Engineers of JAPAN (IEICEJ).

Tomonori Aoyama

Tomonori Aoyama received his B.E., M.E., and Dr. Eng. degrees from Tokyo University, Tokyo, Japan, in 1967, 1969, and 1991, respectively. He joined NTT's Electrical Communication Laboratories in 1969 and worked there on the research and development of transmission systems such as data modems, PCM channel banks, digital multiplexers, digital cross-connect systems, and speech coding, as well as that of digital signal processing for communication systems. From 1973 to 1974, he was at M.I.T. as a visiting scientist to do research on digital signal processing technology.

In recent years he has been executive manager of a laboratory responsible for telecommunication quality, then of a laboratory responsible for broad-band transmission systems, and of the Intellectual Property Department. He is currently the director of NTT Opto-Electronics Laboratories.

Dr. Aoyama is a member of IEEE and IEICE, Japan. He served as a guest editor for the IEEE J-SAC issues on "Voice Coding for Communications," issued in February 1988; on "Packet Speech and Video," issued in June 1989; and on "Intelligent Signal Processing in Communications," issued in December 1994 and January 1995. He is currently a member of the board of directors of the IEICE.

Index

The Artech House Telecommunications Library

Vinton G. Cerf, Series Editor

Advanced Technology for Road Transport: IVHS and ATT, Ian Catling, editor

Advances in Computer Communications and Networking, Wesley W. Chu, editor

Advances in Computer Systems Security, Rein Turn, editor

Advances in Telecommunications Networks, William S. Lee and Derrick C. Brown

Analysis and Synthesis of Logic Systems, Daniel Mange

Asynchronous Transfer Mode Networks: Performance Issues, Raif O. Onvural

ATM Switching Systems, Thomas M. Chen and Stephen S. Liu

A Bibliography of Telecommunications and Socio-Economic Development, Heather E. Hudson

Broadband: Business Services, Technologies, and Strategic Impact, David Wright

Broadband Network Analysis and Design, Daniel Minoli

Broadband Telecommunications Technology, Byeong Lee, Minho Kang, and Jonghee Lee

Cellular Radio: Analog and Digital Systems, Asha Mehrotra

Cellular Radio Systems, D. M. Balston and R. C. V. Macario, editors

Client/Server Computing: Architecture, Applications, and Distributed Systems Management, Bruce Elbert and Bobby Martyna

Codes for Error Control and Synchronization, Djimitri Wiggert

Communications Directory, Manus Egan, editor

The Complete Guide to Buying a Telephone System, Paul Daubitz

Computer Telephone Integration, Rob Walters

The Corporate Cabling Guide, Mark W. McElroy

Corporate Networks: The Strategic Use of Telecommunications, Thomas Valovic

Current Advances in LANs, MANs, and ISDN, B. G. Kim, editor

Digital Cellular Radio, George Calhoun

Digital Hardware Testing: Transistor-Level Fault Modeling and Testing, Rochit Rajsuman, editor

Digital Signal Processing, Murat Kunt

Digital Switching Control Architectures, Giuseppe Fantauzzi

Telecommuting, Osman Eldib and Daniel Minoli

Telephone Company and Cable Television Competition, Stuart N. Brotman

Teletraffic Technologies in ATM Networks, Hiroshi Saito

Terrestrial Digital Microwave Communciations, Ferdo Ivanek, editor

Transmission Networking: SONET and the SDH, Mike Sexton and Andy Reid

Transmission Performance of Evolving Telecommunications Networks, John Gruber and Godfrey Williams

Troposcatter Radio Links, G. Roda

UNIX Internetworking, Uday O. Pabrai

Virtual Networks: A Buyer's Guide, Daniel D. Briere

Voice Processing, Second Edition, Walt Tetschner

Voice Teletraffic System Engineering, James R. Boucher

Wireless Access and the Local Telephone Network, George Calhoun

Wireless Data Networking, Nathan J. Muller

Wireless LAN Systems, A. Santamaría and F. J. López-Hernández

Writing Disaster Recovery Plans for Telecommunications Networks and LANs, Leo A. Wrobel

X Window System User's Guide, Uday O. Pabrai

For further information on these and other Artech House titles, contact:

Artech House
685 Canton Street
Norwood, MA 02062
617-769-9750
Fax: 617-769-6334
Telex: 951-659
email: artech@world.std.com

Artech House
Portland House, Stag Place
London SW1E 5XA England
+44 (0) 171-973-8077
Fax: +44 (0) 171-630-0166
Telex: 951-659
email: bookco@artech.demon.co.uk